CENTRAL HEATING

CENTRAL HEATING

Julian Worthington and David Knight

W. Foulsham & Co. Ltd.

London · New York · Toronto · Cape Town · Sydney

W. Foulsham & Company Limited
Yeovil Road, Slough, Berkshire, SL1 4JH

ISBN 0-572-01328-0

Printed in Spain by Cayfosa. Barcelona
Dep. Leg. B-3813-1986

Contents

Introduction   7
Types of central heating   7
Heating fuels   7
Home insulation   8
Grants   9
Regulations   9
Working with units   9

1   Choosing the system   11
Underfloor heating   11
Ducted warm air   12
Storage radiators   12
Sealed system   13
Small bore system   14
Microbore system   15
Types of fuel   15
Types of control   19
Comparing costs   20

2   Plumbing fittings   22
Pipes   22
Pipe fittings   25
Valves   28

3   Boilers and flues   33
Solid fuel boilers   33
Oil-fired boilers   36
LPG-fired boilers   37
Gas-fired boilers   38
Free-standing boilers   40
Isolated boilers   40
Flues   41
Gas combination units   45

4   The controls   46
A basic system   46
Timeswitches and programmers   48
Thermostats   50
Thermostatic radiator valves   51
Motorised valves   53
Electrical safety   54

5   Planning the system   56
Calculating heat needed   56
Types of radiator   66
Siting radiators   71
Calculating radiator size   75
Designing the pipe circuit   76
Calculating pipe sizes   78
Calculating the pump size   81
Calculating the boiler size   81

6   Installing the system   82
The installation work   88
Lifting floorboards   95
Notching joists   97
Plumbing in   99
Microbore system   101

7   Getting the system working 107
Filling the system   107
Connecting up the controls   108
Commissioning the system   108
Balancing the system   111

Index   114

Introduction

Central heating goes back to Roman times and in the United Kingdom there is still evidence of some remarkably sophisticated systems, dating back more than 2000 years, installed in villas of that period.

It did, however, take a lot longer for the idea of overall heating to catch on and it is only in the last century that this form of heating has become part of our everyday way of life.

As with many other aspects of the home, central heating has now been put within reach of the DIY enthusiast and manufacturers have developed systems that can be installed by people with a reasonable knowledge of plumbing and a degree of ability in DIY.

Cost is obviously one of the determining factors, but prices have now been brought within the range of the average householder and there is no doubt that central heating is a worthwhile investment.

Types of central heating

There are three basic options available for the average home in terms of central heating systems. Depending on the budget available, partial central heating can warm specific rooms in the home. A further compromise is background central heating, which will provide an overall temperature of about 14–16°C, with additional, direct warmth being provided by individual heat sources fired by electricity, gas or solid fuel. With full central heating, each room can be maintained at the required temperature via a single main source of heating. With all these systems, however, you will need a separate heating circuit for domestic hot water.

The controls necessary and available for these different systems vary according to the size and complexity of individual designs. As a general rule, the more complex the controls, the more efficient the system becomes, provided that the various elements are correctly planned and installed.

There are many different ways in which central heating can be provided in the average home. These include underfloor heating, ducted warm air, electric storage radiators and the more conventional 'wet' systems using standard or convector radiators – or both.

Underfloor and ducted systems are usually only installed while the house is being built and are not recommended as DIY systems to existing properties. They are therefore not covered in this book.

Storage radiators that use off-peak electricity have been available for many years and are now much more streamlined and efficient. The current range incorporates many features not available with earlier fittings. Your local Electricity Board will be pleased to supply all the information you need on this type of domestic heating.

In this book we have concentrated on the various types of 'wet' system that can be installed in your home, whether fuelled by electricity, gas, oil or solid fuel.

Heating fuels

Of the range of fuels available for 'wet' systems, the traditional form is solid fuel, which is, of course, still available. Modern boilers have come a long way towards removing the chores of cleaning away ashes and refuelling and the current hopper-fed models only need attention once a day – and sometimes less. Although solid fuel can be delivered to any home reasonably accessible by road, sufficient storage space is necessary and desirable if advantage is to be taken of the lower summer fuel prices.

Gas is probably the most popular choice for central heating systems, since it is clean, instantly controllable and still relatively cheap, despite recent price increases. The only restriction is whether piped gas is available in your particular area. If it is, but you do not have gas laid on, your local Gas Board will advise on how to have your house connected and how much it will cost.

If you are unfortunate enough to live in an area where piped gas is not supplied, you do have an alternative in bottled gas, which can either be provided in exchangeable cylinders or stored in refillable tanks.

Although oil is now not as popular as it used to be, due to substantial increases in price a few years ago, it still provides a viable alternative to gas where this is not available. Like gas, it is relatively clean and instantly controllable. You will, however, need to have a large enough storage tank – which, incidentally, must be accessible from the road – to take advantage of bulk delivery discounts.

Electricity, although until recently not considered as economical as gas, has made up a lot of ground and must now be considered very competitive as a source of fuel, although it is not as easily controllable. This is because to make it economically viable, it is necessary to take advantage of the off-peak Economy 7 tariff. This involves heating during the night time and utilising this stored heat during the day.

It is difficult to assess accurately the relative costs of different types of fuel. Fluctuations in price are one factor to bear in mind. The type of system and what your requirements are will also be relevant. And there are other related factors such as the effectiveness of existing insulation. Advice on heating cost comparisons can be gained from the Department of Energy's leaflet *A Guide to Home Heating Costs*.

Home insulation

The effect of central heating – whatever the system you decide to install – will to a large extent depend on how well your house is insulated. There is little point in spending what will be a considerable amount of money on ensuring good overall heating for your home if you do not take the necessary precautions against losing that heat.

Heat, of course, is always lost eventually. But the point of having adequate insulation is that you can delay the inevitable for as long as possible and therefore take maximum advantage of what heat is there while it is retained. You are therefore strongly recommended to read *Home Insulation* (published by W. Foulsham), which is included in this *Know How* series, to check on what forms of insulation are available and need to be fitted.

As a general guide, the following precautions – in order of importance – should be taken to protect and maximise your investment in a central heating system:

Hot water cylinder   This should be covered with an insulation jacket to BS 5615:1978 which is 75 mm (3 in) thick. The price of this jacket can be recouped in a matter of weeks in terms of fuel costs saved.

Loft insulation   Make sure your loft has at least 100 mm (4 in) of floor insulation between the joists if using a blanket material or 150 mm (6 in) if using loose-fill. Grants are available for this insulation, the costs of which can, in any case , be recovered in 3–4 years.

Wall insulation   If your house is constructed with cavity walls these should be insulated with polyurethane foam, polystyrene beads or mineral fibre. This job must be carried out by a qualified specialist, the cost of which can be recovered in 5–7 years.

Draughtproofing   It has been estimated that the draughts present in the average house can be represented by a hole one metre square in an external wall. This type of insulation – in windows, doors and floors – is relatively cheap and easy to install and for that reason is very cost-effective. However, you must make sure that you still allow sufficient air to circulate through the house for the purposes of ventilation.

Double glazing Single glazed windows are poor insulators and a large amount of internal heat is lost through them. Equally they are a major source of draughts. Double glazing, if installed by a contractor, is not cheap but it does reduce heat loss, eliminate draughts and condensation and cut down on noise. DIY double glazing is much cheaper and well within the capabilities of the home handyman, as shown in the companion volume in this series – *Home Insulation*.

Grants

As mentioned above, local authority grants are available for loft insulation, if none is present or if it is less than 30 mm (1¼ in) deep, to ensure a thickness of 100 mm (4 in). If the grant is approved, it will cover 66 per cent of the material costs and labour charges. This means that it is possible to have your loft insulated professionally at a price only slightly more than it would cost you to do it yourself.

The local authority may also make discretionary grants towards the installation of central heating in certain cases. Your local Citizens Advice Bureau will advise on whether it is worth applying for one of these in view of your own particular circumstances.

Regulations

The planning and water authorities do have varying regulations depending on the area in which you live. These include such things as the suitability of different fittings – for example, in some soft water areas brass fittings are prohibited – and which appliances can be fed direct from the mains water supply and which must be connected to the cold water storage cistern. Equally, you may find that an external flue for the boiler requires planning permission.

Severe penalties can be incurred if these regulations are ignored and so it is in your own interest to check that whatever work or installation you are planning does not contravene the existing regulations.

It is common sense, but a point worth stressing, that any final gas connections should be made by a qualified engineer and any electrical wiring should conform to the latest IEE regulations. These can be checked via your local library.

Equally it is worth keeping up to date with British Standards and the relevant codes of practice are BS CP 331 and 342. The other British Standards that may prove useful are: BS 5376 Pt 2, BS 5440 Pt 1 & 2, BS 5446 and BS 5449 Pt 1.

Working with units

Where applicable, conversions of measurements are given in this book; otherwise all measurements are in metric. This is simply because plumbing fittings are now produced in such units and are not convertible into Imperial.

For general interest and to enable conversions to be made where necessary, a set of conversion factors is given below. These conversions have been rounded off to give you a sufficiently accurate estimate without getting bogged down in cumbersome figures:

length:	1 in = 25.4 mm
	1 m = 1000 mm = 3.28 ft
volume:	1 cubic metre (m^3) = 35.3 cu ft
capacity:	1 litre = 0.22 gallons
	1 gallon = 4.54 litres
pressure:	1 Newton/m^2 (Pascal)
	= 0.1 mm water gauge (WG)
	= 0.004 in water gauge (WG)
	1 millibar = 100 Newton/m^2
	= 10.24 mm WG
	= 0.403 in WG
	= 0.0145 PSI (lb/sq in)

temperature: $-1°C = 30°F$
                                 $0°C = 32°F$
                                 $16°C = 61°F$
                                 $18°C = 64°F$
                                 $21°C = 70°F$

heat: 1 kilojoule (kJ) = 1000 joules
                       = 0.948 BTU (British Thermal Units)
1 BTU = 1.055 kJ
1 watt = 1 joule/sec = 3.412 BTU/hr

1   Choosing the system

When it comes to installing central heating in the home, there is a range of systems available. At the end of the day, the choice is yours, although there will be factors that will inevitably influence this choice.

You may decide on the basis of the cost comparisons of different types of central heating. Equally, if you wish to do some – or all – of the work yourself, this will definitely limit your choice.

Other relevant factors will be the availability of fuel since, for example, some areas are not connected to mains gas supplies, or problems of storage, such as you could have with oil or solid fuel.

Here are the basic types of central heating available for domestic situations, with some comments on relevant advantages and disadvantages.

Underfloor heating

With this system, the solid floors themselves are heated, nowadays normally through electrical elements built into the concrete. The idea with underfloor heating is to make use of off-peak electricity during the night, with the floor giving off its heat during the day.

Naturally this system can only practically be installed when the house is being built. But there are other disadvantages as well. The heat is not easily controllable, since, once the floor has been heated, you cannot then stop the heat if there happens to be a sudden rise in the overall temperature.

Another potential snag with this type of heating arises if you want to fit thick pile carpets on the floor. The thicker the carpet, the greater the insulation barrier you create and therefore the less efficient the heating becomes. You must also bear in mind that heat will not only travel upwards but also downwards, so that some of the heat you are paying for is going down

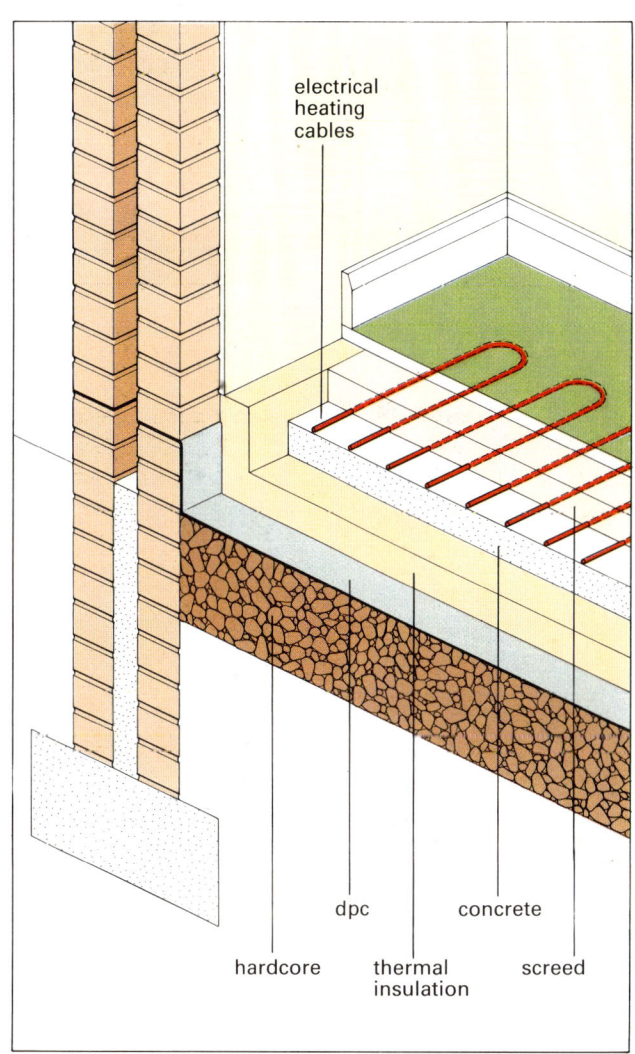

This is the most common form of underfloor heating. Electric coils, often built into the concrete, heat the screed while thermal insulation helps prevent the loss of heat downwards.

and not up, where you want it.

A further point to bear in mind is that you will need a separate system to provide domestic hot water.

Ducted warm air

This system works on the principle of warm filtered air being blown along ducts in the walls and under the floors and discharged into the room through grilles fitted at floor level. At the same time any stale air in the room is removed through other grilles fitted higher up on the walls.

There are two methods used to heat the air. One involves a 'dry' heat exchanger using gas, oil or electricity, where the air is passed over heated elements. The other involves a 'wet' heat exchanger where the air is passed through a radiator heated with hot water. In this case, you can use the same system to heat the domestic hot water as well.

With the more elaborate – and expensive – systems, you do have the facility to circulate cooled air during the summer.

One great advantage of this type of central heating is that it is instantly controllable. Unfortunately, however, it is not completely silent in operation. Although it can be installed into an existing house, the ducting required is quite bulky and rather unsightly. Normally, therefore, it would be installed when the house is being built.

Storage radiators

Central heating using storage radiators is probably the easiest system to install. You simply have fitted the radiators you require in the rooms through the house. These radiators comprise of heating elements contained in heat-retaining blocks inside the casing.

As with underfloor heating, off-peak electricity is

A modern storage radiator with controls situated on the top of the heater. The output is controlled by a fan to achieve maximum efficiency.

used. This means the elements are switched on during the night when they heat up the blocks. The stored heat is then released into the room during the day. You can have a second, shorter heating period around the middle of the day which acts as a top-up to maintain a reasonable heat level through the evening.

Earlier models offered limited temperature control, since the heat built up during the night was simply released gradually during the day. Latest models incorporate a control which enables you to vary the amount of heat you want to store and a separate thermostat control operating a damper to vary the rate at which heat is emitted during the day.

The advantage of these controls is obvious, since they allow you to maintain a relatively even temperature in the room over a 24-hour period. If you already have the older type of storage heater, it is sometimes possible to have it adapted with a thermostatic control. Check with your local Electricity Board if you want to alter an existing fitting.

If you have storage heaters fitted, the electricity for them is supplied on a cheaper tariff known as Economy 7. The Electricity Board fits a special timeswitch and dual-reading meter and the electricity is supplied during seven hours of off-peak demand – in most cases between midnight and 8am, although this does vary slightly from one area to another. This electricity is charged at a special low rate – less than half the standard domestic tariff.

During this low-tariff period, it is possible for you to make use of the cut-price electricity to use other high-consumption appliances such as immersion heaters, washing machines and tumble dryers.

In the day time, your electricity is charged at a slightly higher rate than the normal domestic tariff and your standing charge is also higher.

This system is the easiest to install because all it requires is a special electrical circuit to feed your storage heaters, which are permanently wired in. Because these heaters are very heavy, it may in some cases be necessary to strengthen the floor area underneath.

If you are considering having this type of central heating installed, contact your local Electricity Board, who will carry out all the necessary heat calculations and provide an estimate for the installation work.

Sealed system

This is what is known as a 'wet' system, using a boiler and radiators, in which the circuit that heats the radiators and the calorifier in the hot water storage cylinder is sealed under pressure.

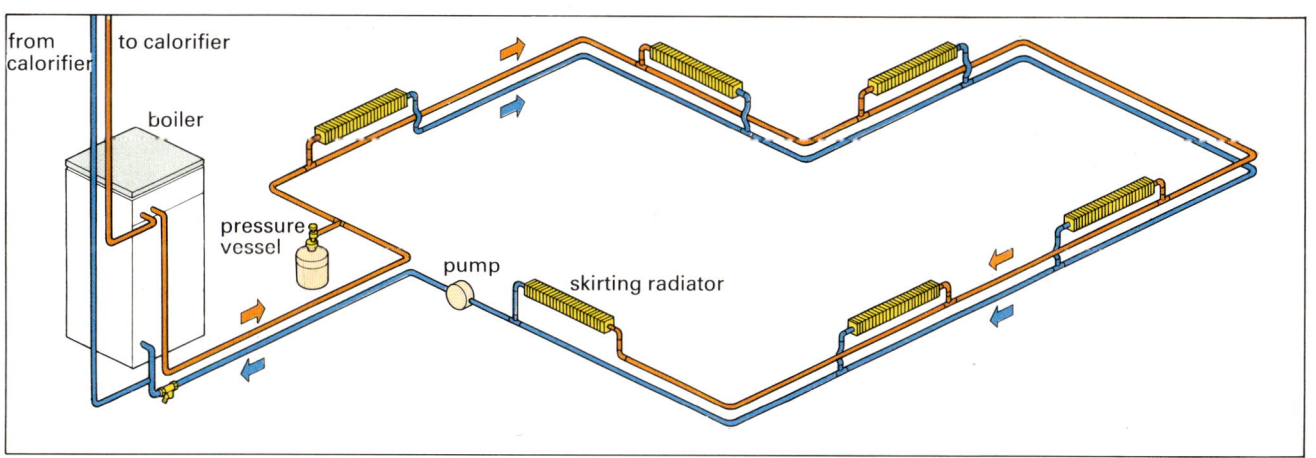

This diagram illustrates how a basic sealed system works. Here there is no feed and expansion tanks and the difference in the volume of water caused by heating is taken up by the pressure vessel. Normally this type of system operates at a higher temperature than a normal central heating system and uses skirting or convector radiators. As the system does run at a higher temperature, it is important that all pipework is covered.

This type of system is used in buildings where the header tank cannot be situated at a sufficient height (for example, in a flat) when temperatures of less than 80°C are used throughout the system. The other, more common use is where the system is designed to work near to or above the boiling point of water; water boils at higher temperatures when under pressure. In this case, all pipework must be covered and a fan convector or skirting radiators may be used, since they have no exposed parts at water temperature.

Only one header tank for the domestic hot water is needed and this can be situated just below ceiling level. A pressure vessel, charged with an inert gas (usually nitrogen) is fitted to compensate for the difference in volume of the water in the heating circuit from when it is cold to when it is hot.

If properly designed and installed, this type of system is as safe as any other. Obviously there must not be a leakage of water from the heating system and so stringent safety regulations are applied. In certain cases you can encounter some opposition to the system from local water authorities and property insurance companies.

As a rule, the sealed system is an expensive one to install. For this reason – and the fact that it has a specialised application – it has not been covered in this book.

Small bore system

This is the most common system used in domestic central heating. The small bore referred to is the 15 mm bore piping normally used to feed the radiators in the system – as opposed to the $\frac{3}{4}$ in and 1 in steel pipes that used to be fitted.

The system starts at the top of the house with a small feed and expansion tank fitted in the loft. This tank supplies the water used in the primary circuit. This circuit feeds the boiler, radiators and the

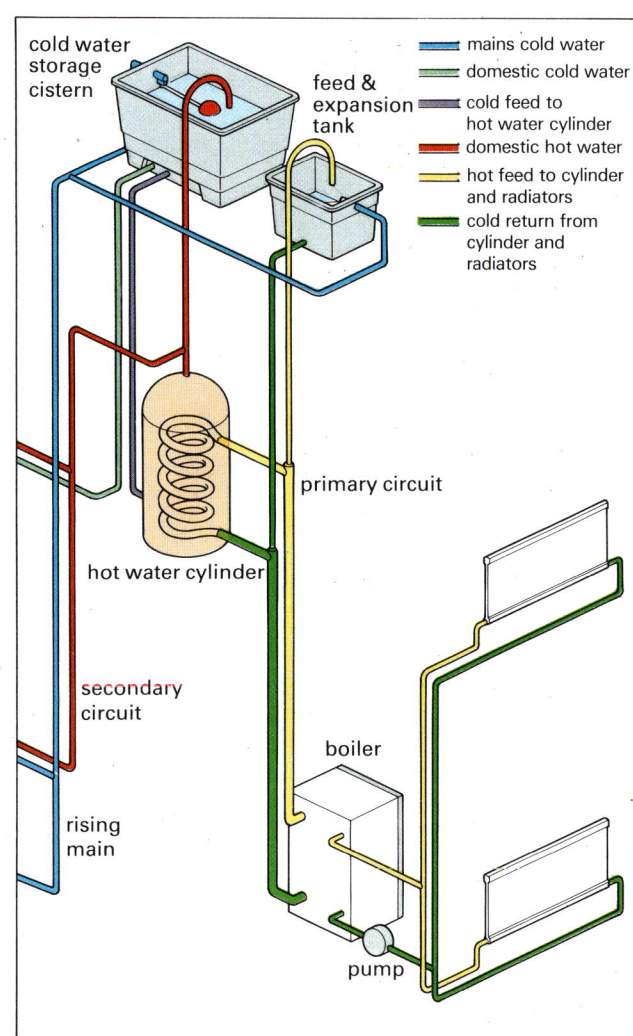

Here you can see the basic small bore system which is commonly found in many homes. The system combines a pumped central heating circuit with a gravity-fed domestic hot water circuit.

calorifier in the hot water cylinder. The water in this circuit is not drawn off, but remains continually in the circuit. The reason for this is to eliminate the possibility of corrosion and scale formation, which were the main problems with the original direct system.

From the feed and expansion tank a pipe runs down to the bottom of the boiler. The return pipe from the calorifier is also connected to this down pipe. From the top of the boiler a vent pipe returns to the loft, culminating in a U bend over the top of the feed and expansion tank. The feed to the calorifier is normally taken from this vent pipe.

There are two ways the water in the circuit can be fed through the calorifier. This can either be done by gravity or through the use of a pump (see pages 108–113).

From the top of the boiler a feed pipe is taken to the radiator circuit and the return from the radiators is fed into the bottom of the boiler. At some point in this circuit a pump is fitted to move the water around the circuit and through all the radiators.

The domestic hot water circuit incorporated into this system is known as the secondary circuit. It starts from the cold water storage cistern in the loft, from where a cold feed pipe is taken normally to all cold water outlets in the house, except the kitchen sink. Here the water is supplied direct from the mains.

A second pipe runs down from the cold water storage cistern to the bottom of the hot water cylinder. The cold water then surrounds the calorifier. Having been heated by the water from the boiler, the calorifier in turn heats up this secondary water in the cylinder.

From the top of this cylinder, a vent pipe runs back up to the loft, terminating in a U bend over the cold water storage cistern. The supply to the hot water taps and outlets in the house is taken from this vent pipe.

As a general rule, this system is the easiest for you to install yourself. Pipes and fittings are readily available and the complete installation in an existing house can be carried out with the minimum of disruption.

In large houses with long pipe runs, you may have to use pipes of a larger diameter in order to maintain the recommended flow of water. While 15 and 22 mm copper pipe can be manipulated fairly easily by hand, the larger sizes will require the use of a bending machine.

Microbore system

This is a fairly recent innovation to central heating, based on the small bore system and using similar installation principles. The main advantage of this system is that the pipes used are much smaller (6, 8 and 10 mm bore) and therefore much easier to route and conceal as they are fed round the house.

It is sometimes possible to work from a basic small bore system and reduce the size of the feed pipes to the radiators, but this is not commonly done. More frequently the hot water is fed through small bore pipes to manifolds on each floor level and from these manifolds microbore pipes are used to feed the radiators. The calorifier may also be fed using small bore or microbore pipes.

Types of fuel

There are basically four types of fuel that can be used with the range of central heating systems already mentioned. You should consider carefully the practicalities and costs of each type, since this will be important when making a final decision on the particular system you want installed.

Solid fuel The traditional style boiler was always run on solid fuel and this type is still available, normally fired with coal, anthracite or coke. You should check

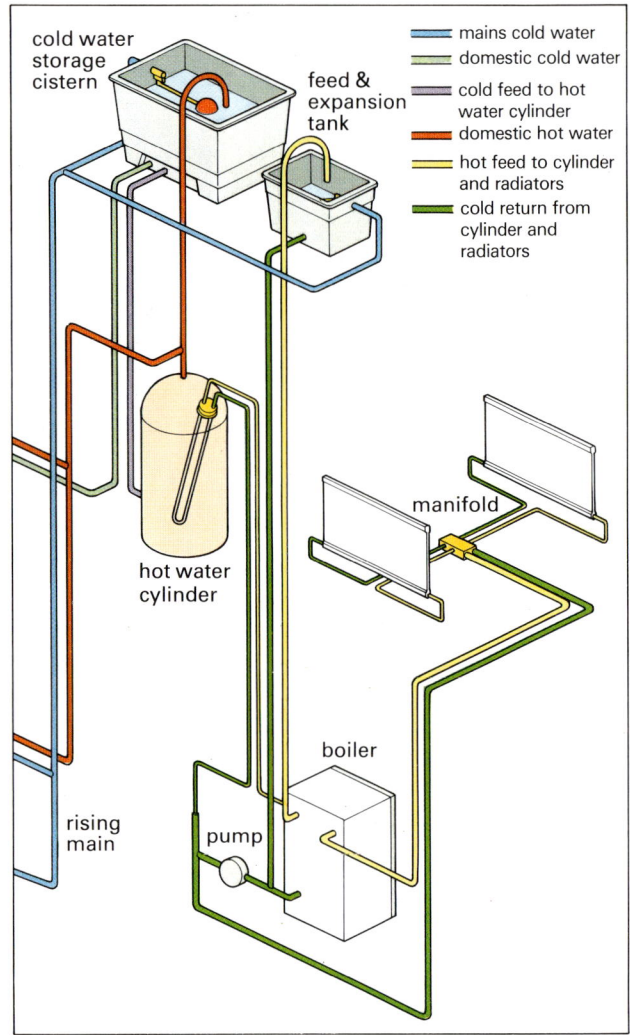

cold water
storage
cistern

feed &
expansion
tank

mains cold water

domestic cold water

cold feed to hot
water cylinder

domestic hot water

hot feed to cylinder
and radiators

cold return from
cylinder and
radiators

hot water
cylinder

manifold

boiler

rising
main

pump

This typical microbore system shows how the hot water is
supplied to the radiators via special manifolds. The hot
water is pumped round the central heating system and the
domestic hot water system as well.

on the local smoke control regulations since in many areas you will be limited to the use of either anthracite or coke.

There are several points about solid fuel that you should bear in mind if you want to use this type of fuel. It must be kept in a dry store sufficiently spacious to hold substantial stocks; the Coal Utilisation Council recommends a minimum of 2 tons. This will enable you to take advantage of the lower prices that normally apply during the summer period. The store should be not more than 45 m (150 ft) from the road and there should be reasonable access to allow sacks to be carried easily – the minimum space should be 2.3 m ($7\frac{1}{2}$ ft) high and 1 m ($3\frac{1}{2}$ ft) wide.

With solid fuel boilers you will not be able to control the heat instantly and you will need to have some means of using up the surplus heat generated by the boiler. This normally takes the form of a gravity-fed calorifier, with which to heat the domestic hot water.

Modern boilers fall into two basic categories. The first involves a room heater, which fits into a fireplace and has a back boiler to heat the central heating and domestic hot water. This type of heater is normally fitted with glass doors and is fuelled manually, although automatic hopper-fed models are available. The heat control for this type of boiler is normally by means of a thermostatic damper.

The other type is free-standing, nowadays normally hopper-fed and gives off little radiant heat. Again the heat is generally controlled by a thermostatic damper, although the better models have an electric fan that offers a much quicker response and superior control.

Both types of boiler do require some daily attention, since the ash or clinker has to be removed. You will also have to have the chimney and flue cleaned periodically, depending on the type of fuel used.

You can get wood-burning stoves to feed central

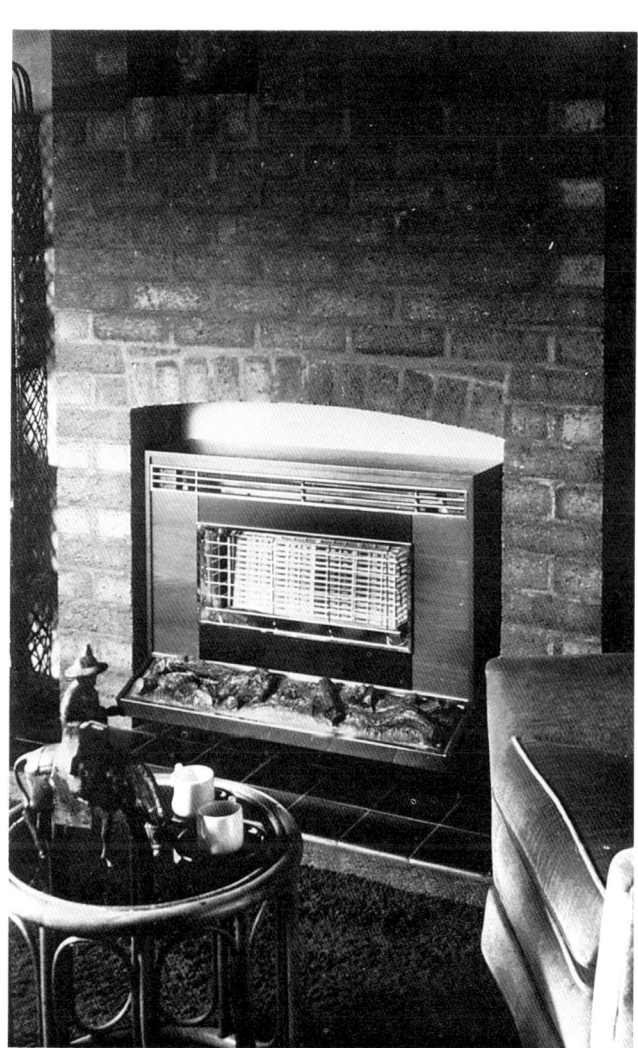

This modern design solid fuel room heater not only provides an attractive feature in a living room, offering direct heat, but has the additional advantage of heating water as a boiler at the same time.

Solid fuel boilers have been streamlined in recent years to overcome many of the original drawbacks involved with using this type of fuel. This hopper-fed model, with thermostatically controlled damper, fits in any kitchen.

heating systems, but the main problem here is that you would have to ensure a regular supply of suitable wood. For this reason you may decide that this type of boiler would not be a feasible alternative.

Oil   This used to be a popular choice of fuel for central heating systems until the price increases in the 1970s made it a relatively expensive method of heating. But you may find it is a viable alternative where, for example, there is no mains gas supply.

One advantage of oil-fired boilers is that they are instantly controllable. Also oil can be supplied to any area, although you will need a large tank in which to store it. The recommended minimum size for domestic use is 2700 litres (600 gallons) capacity, which allows for deliveries of 2275 litres (500 gallons) at a time.

Oil-fired boilers are self-contained, but not completely silent in operation, which is why they are usually fitted away from the normal living quarters of the house. Most have conventional flues, which require no attention once they are installed.

To ensure efficient operation at all times, oil-fired boilers do need to be serviced regularly.

Gas   This is probably the most common form of fuel used in central heating systems, the main reasons being that gas is relatively economical, instantly controllable and clean. Although some routine maintenance is necessary to ensure efficient operation, no day-to-day attention is required since gas-fired boilers are automatically controlled. For safety reasons, control devices fitted to these boilers prevent the emission of gas in the unlikely event of a flame going out or failing to ignite.

Mains supply gas is not available to approximately 15 per cent of the homes in the United Kingdom – and in this case you will have to consider the use of propane gas or choose an alternative form of fuel. Propane gas does work out more expensive than mains gas and you will need to buy and install a suitable supply cylinder.

If there is gas in the area but you are not yet supplied, connection is relatively inexpensive, provided that your property is within 23 m (75 ft) of an existing gas mains.

It is reasonably easy to install a gas boiler. In many cases this can take the form of a room sealed boiler fitted with a balanced flue. With this arrangement, no extra ventilation is required to supply air for combustion, which has the additional advantage of eliminating a source of draughts in that room – and therefore potential waste of heat. Where it is not possible to fit this type of gas boiler, then a normal flue will be necessary.

Electricity   One use of electricity has already been discussed in connection with storage radiators (see pages 12–13). It can also be used in the form of radiant heaters, convector heaters and oil-filled radiators. Heating a whole house with these appliances is, however, relatively expensive and normally you should only use them to boost the temperature rather than provide background heat.

Electricity can be used to fuel a central heating system through an Economy 7 boiler. Although this is primarily designed to replace an oil-fired boiler, it is suitable when planning a new system.

The boiler is basically a large, well-insulated water storage vessel with immersion heaters fitted through the side. A typical domestic model would have a capacity of approximately 675 litres (150 gallons). In some cases, heaters are fitted to the top of the boiler to provide hot water direct to the radiators should the stored water not be sufficiently hot (see diagram).

The heaters normally operate during the night using off-peak electricity. When the stored water reaches the required temperature (95°C), they cut off, since they are thermostatically controlled. A normal control system switches the pump on and off as required, feeding the hot water through the radiators

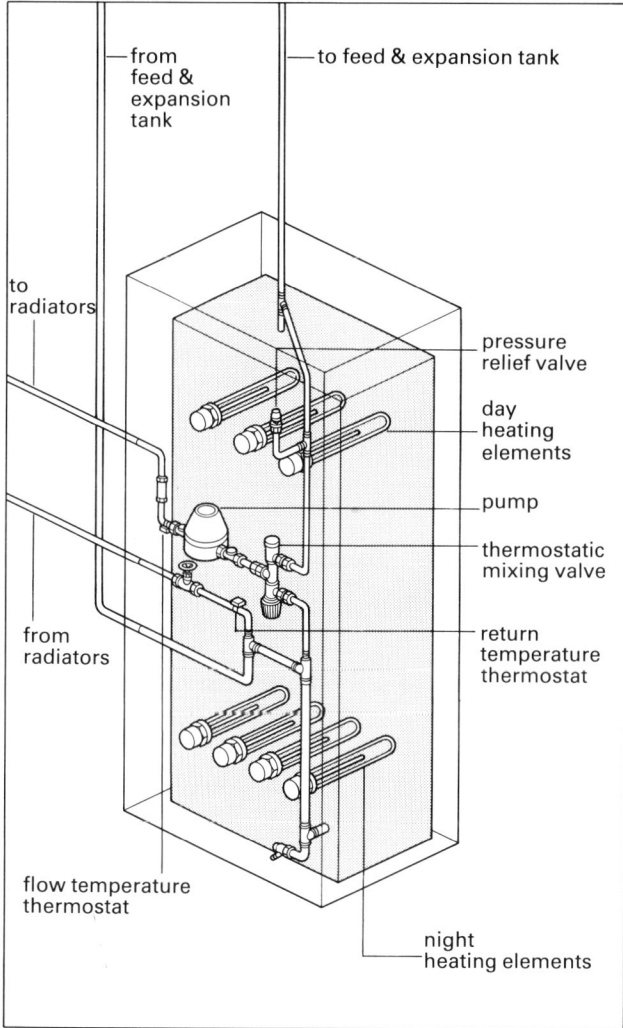

from
feed &
expansion
tank

to feed & expansion tank

to
radiators

pressure
relief valve

day
heating
elements

pump

thermostatic
mixing valve

return
temperature
thermostat

from
radiators

flow temperature
thermostat

night
heating elements

This diagram shows how an Economy 7 boiler works. The boiler contains a large volume of water that is heated by off-peak electricity. To top up the level of heat, the day elements in the boiler can be used.

during the day.

The stored hot water is gradually used up during the day, before it is reheated at night. You can, of course, override the controls and switch on the immersion heaters to boost the temperature when required. In this situation, however, you will be using electricity at the full rate.

The Economy 7 boiler can be used to heat the domestic water as well – or you can use the more traditional methods such as a separately controlled immersion heater in the hot water storage cylinder. With this system, incidentally, it is best to have a dual-element type. This enables you to heat just the top part of the cylinder for normal washing requirements and the rest of the cylinder when you need the water for a bath.

Your local Electricity Board will happily supply details of this type of boiler. Before installation, check that the floor is strong enough to carry the weight, since the average boiler will contain anything up to 850 litres (190 gallons) of water.

In very cold weather or during sudden chilly spells, the stored heat will be used up during the day. This means full tariff electricity may well be needed to heat up the radiators during the evening, which can prove relatively expensive.

Types of control

From the point of view of fuel economy, it is very important that the boiler does not run for longer than necessary to provide hot water to heat the house to the desired temperature. One of the main controls, therefore, will be a timeswitch. This enables you to have the boiler on at times when the heating is mainly required and, equally, off when heating is not needed.

Most timeswitches have facilities for two separate heating periods – and some for three – during each 24 hours and also for domestic hot water or central

For fuel economy insulating jackets should be fitted to hot water cylinders and tied into place with straps. Warmth will still filter through into the airing cupboard.

heating – or both. The settings can easily be altered to suit the requirements of the household.

You will want some form of thermostatic control to ensure that when specific areas of the house reach the desired temperature the boiler will automatically cut out and then switch on again when the temperature drops below that level.

Where you position the thermostatic control switches is a matter of choice, although you do not want to put them in rooms where there are liable to be dramatic changes in temperature, such as the bathroom or kitchen. Another effective method of controlling the temperature is the use of thermostatic valves on each radiator.

If the house you live in is in a particularly exposed situation or is likely to be left empty for a period of time through the winter, it makes sense to include a frost thermostat. This will turn on the boiler when the temperature drops below a certain level to ensure the plumbing does not suffer from frost damage.

The range of controls (see pages 46–54) is now extensive and it is worth checking to see which will be of particular value. As a general rule, the more you have fitted the more efficient your central heating will be at the lowest cost in terms of fuel. Equally you should bear in mind that the more complicated you make the system, the more there is that can go wrong.

Comparing costs

It is, of course, impossible to provide hard-and-fast rules about the relative costs of different forms of central heating. The price of fuels varies from year to year and the picture is confused even further by the periodical increases in cost.

It would appear, however, that oil prices are likely to remain high for the foreseeable future and therefore an oil-fired system, though having much to recommend it, will tend to be more expensive to run than the other basic types. There has also been a levelling off of gas and electricity prices, where in the past gas was certainly the cheapest form of fuel.

The only way to get anywhere near an accurate estimate of likely heating costs for different types of system is to work out the total heat load for the house in watts (see pages 61–66) and then to check with the manufacturers of each system what amount of fuel would be required to supply that amount of heat. Using current prices, you can then make a comparison between the different fuels.

You should, however, consider other factors in this calculation, such as the additional expenses incurred. These might include the cost of having gas connected to the house or the installation of a coal bunker or oil tank.

You may, of course, not be in a position to choose solely on the basis of costs, since your location may dictate to some extent what type of fuel you can have. But it is worth making a comparison if you can and not just taking for granted what you are told about each system.

It is not the job of this book to make individual recommendations. The systems we have looked at are those that are most common and the simplest to install – that is, the small bore and microbore systems with a domestic hot water supply included.

2 Plumbing fittings

You will find that the materials required to install your own central heating should be readily available from a plumbers merchant or good DIY store or can be ordered. As far as tools are concerned, some you may already have. Others you may well feel it is worth buying, although the specialist tools can be hired for the job.

Pipes

The most common type of pipe used in domestic central heating systems is made of copper to BS 2871 Pt 1. This is readily available and reasonably priced. Alternatively you can use stainless steel – to BS 4127 – but this type of pipe is expensive and difficult to bend. Another important point with stainless steel is that you cannot use capillary fittings with it.

The standard sizes of pipe used with a small bore system (see pages 76–79) are 15 and 22 mm diameter. If you are installing a gravity-fed system, you will need pipe of a larger diameter – namely 28 and 35 mm. These sizes of pipe are much more difficult to bend.

Check on the type of pipe used in the existing water system. If it is of galvanised steel, you should consult your local water authority before using copper pipe with it. Under certain conditions, corrosion can occur as a result of chemical reaction between the two metals and you may find you have to take special measures and use isolating fittings as well.

One problem you may come across in houses that are a few years old is that the copper pipes used may be Imperial sizes. These are not fully interchangeable with metric equivalents and, since Imperial size pipe is no longer available, you will have to incorporate special reduction fittings in certain places where you are plumbing in to existing pipework.

When cutting copper or stainless steel pipe to length, you should always use a rotary pipe cutter. Although you can cut pipe with a hacksaw, it is difficult to ensure that you are cutting the pipe square. Another problem with this method is that it creates more swarf which could be washed into the system later on.

Bending pipes Inevitably when installing pipework there will be plenty of corners to be turned. Wherever possible in these situations the pipe should be bent rather than putting in an elbow fitting. There are two basic reasons for this. Firstly water will flow much more freely through a pipe than a fitting and secondly it is much cheaper and usually quicker to bend a pipe than to use a fitting.

Equally you should resist the temptation to use the convoluted tap connectors available in place of bends, since these have a considerable resistance to water flow.

You will find that 15 mm copper pipe will bend quite easily across the knee and you can use the same method for 22 mm pipe, although you will be required to use more strength and will probably first have to anneal – or soften – the metal where the bend is to be formed. Alternatively you can use a bending machine, which can be hired quite cheaply. Not only does the machine make the job a lot easier, but it also ensures an accurate and consistent bend in the pipe. You will also need two bending springs – one for each diameter of pipe. The spring must be inserted where the bend is to be made to prevent the pipe from kinking or creasing.

Before inserting the spring and bending the pipe, make sure the spring is well greased with petroleum jelly. Not only will it be easier to get it in and out of the pipe, but it will also prevent the spring from rusting and getting stuck in the pipe. The easiest way to remove the spring is to overbend the pipe slightly

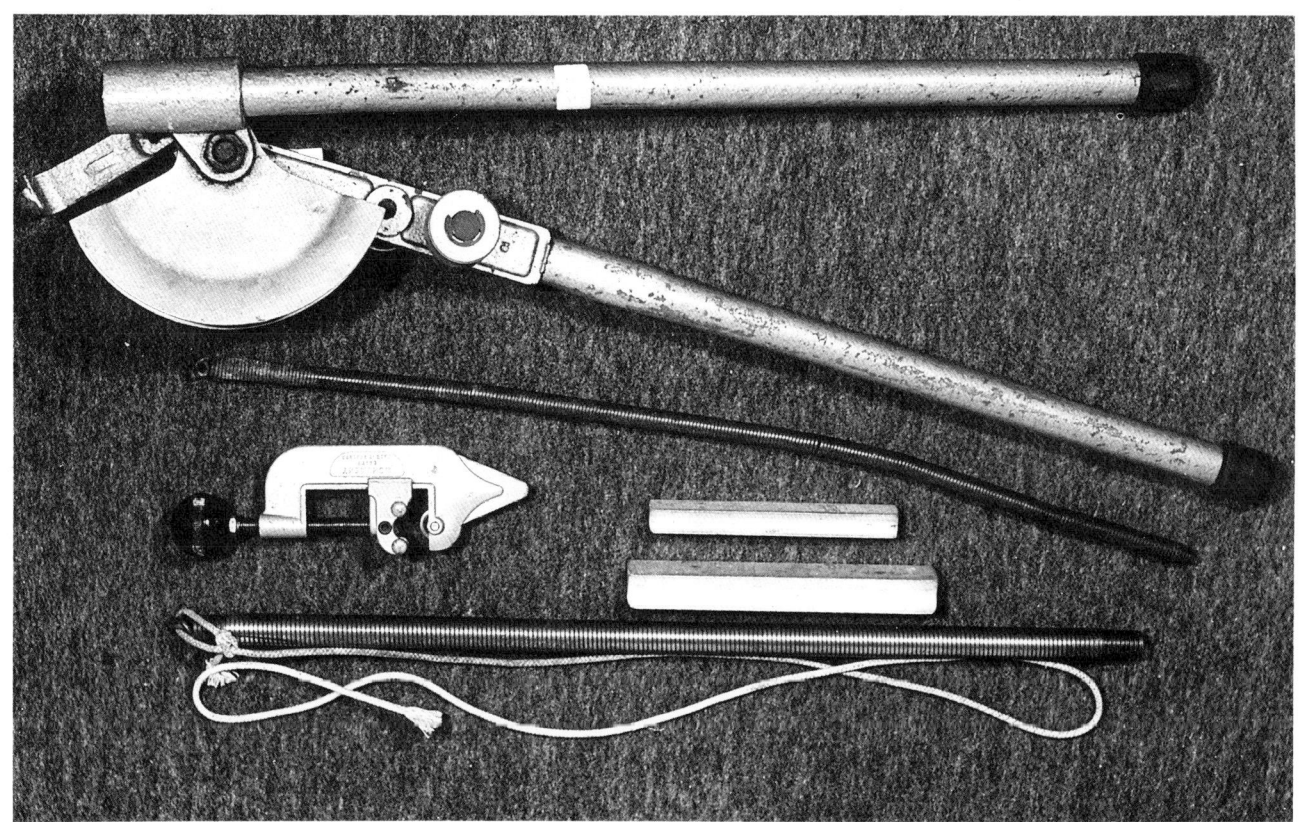

This selection of tools will be necessary for anyone wanting to cut and bend pipe. 1. Pipe bending machine. 2. 15 mm bending spring. 3. Pipe cutter. 4. 15 mm and 20 mm formers for use with the bending machine.

5. 20 mm banding spring. Depending on the amount of work you carry out, you may find it cheaper to hire these tools.

then ease it back to the required angle. This will help release the pipe's grip on the spring.

Sometimes you may have problems getting the bending spring out of the pipe. You can often free it by inserting a small screwdriver into the loop at the end of the spring and twisting the spring clockwise as you pull on it. As a last resort, try tapping all round the

pipe where the spring is jammed with a wooden mallet to free it. This should not be necessary, however, if you bend the pipe correctly.

As already mentioned, if you have any difficulty in bending the copper pipe, you may have to anneal the copper first. If you have already inserted the spring, remove it by twisting it clockwise and pulling the loop

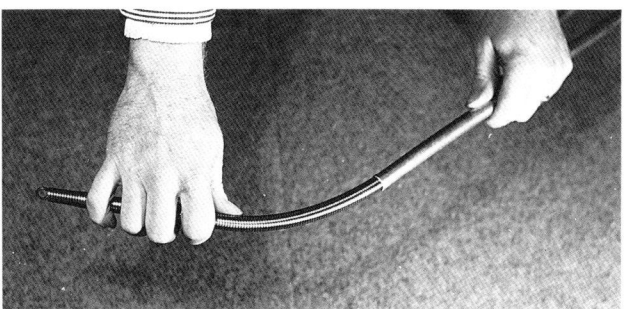

To bend pipe, first insert the greased spring so it is within the section to be bent.

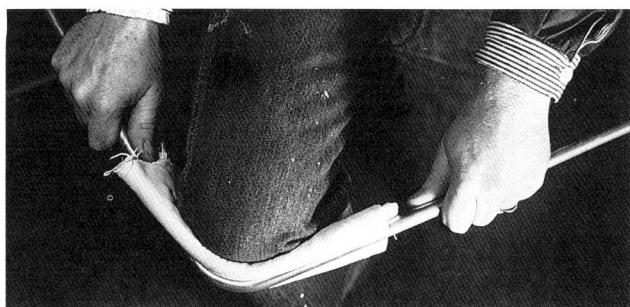

Pull both ends towards you evenly and move the pipe to either side to increase the radius.

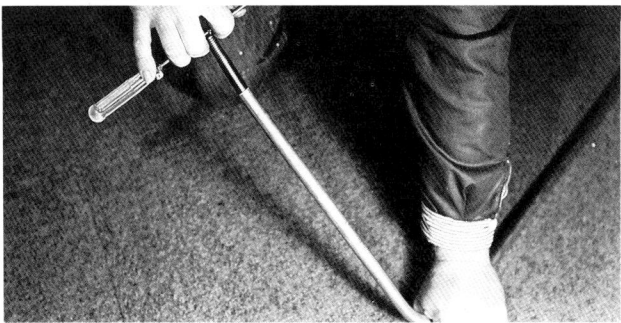

Slightly overbend then adjust the pipe before twisting the spring clockwise and pulling it out.

With a bending machine, insert the pipe with the former in place and pull on the handles.

at the end towards you. Heat the area to be bent with a blowtorch until it glows red and then immediately immerse it in cold water.

The biggest problem when bending pipe is to get the bend in the right position. If you only need one bend in a length of pipe, make sure you allow extra pipe at either end so that you can trim of any excess as required once you have checked the bend on the pipe run.

If you have to make two bends in one length, try where possible to keep them a reasonable distance apart. If they are too close together, you will have trouble making the bends and removing the spring.

If you have to have two bends close together, use separate lengths and join the pipes between the two bends, which will be near the end of each length of pipe. Having made the first bend, work out where the next bend needs to start by measuring up where the pipe run will go and mark this position on the pipe with chalk or a piece of masking tape. Use this marker when making the second bend, ensuring that this is made from exactly that point on the pipe.

It is possible, though not really advisable, to straighten out a bent piece of pipe and use it again. If you have to do this, anneal the bent section of pipe as already described (see above).

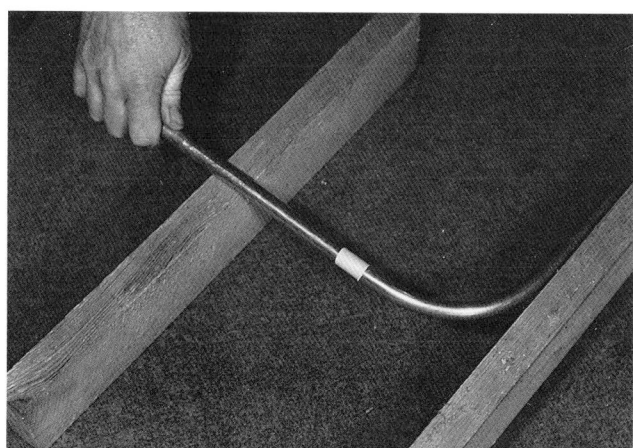

Measure the length of pipe from the start to the finish of the first bend to be made. Mark on the pipe where the second bend will start, measuring from the end of the first bend.

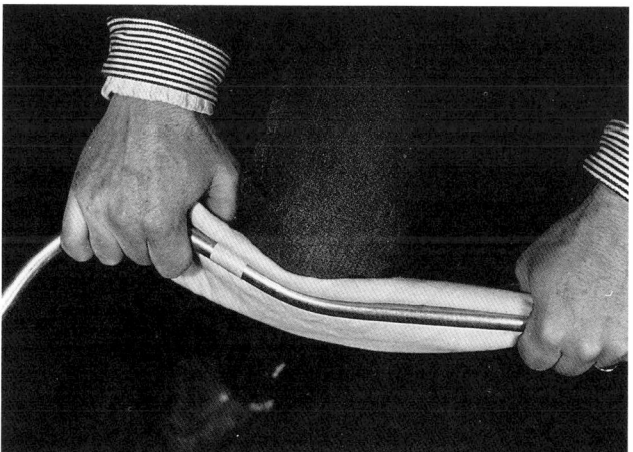

Having measured the length of the second bend, you can make this, working from the middle of this length. It is important to ensure you have enough pipe after the bend to pull on.

If you have to bend a pipe a long way from either end, secure a length of strong cord to the bending spring and measure how far into the pipe the spring will have to go. Mark this length on the cord with a piece of masking tape so you know how far in to insert the spring. You may need a length of thin rod or dowel to push the spring along the pipe. When the bend has been made, pull on the cord to remove the spring. You may need to wrap the end of the cord round a small bar or screwdriver to get enough purchase to extract the spring.

Holding pipes   The changes in temperature in the pipework will cause pipes to expand and contract. If they are held too firmly, this expansion and contraction will cause the pipes to move in their fittings. The clicking noises you may hear in a system when it is cooling down or heating up are a result of this.

You can overcome this by holding the pipe in place with the modern plastic clips and so in a normal house the expansion and contraction of pipes can be ignored.

For the same reason it is not a good idea to embed copper pipe in a concrete floor or behind a plastered wall surface. If you have to run a pipe in these circumstances, you should cut a channel with a separate wooden or metal cover – which should not rest on the pipe (see diagram).

Pipe fittings

There are two basic types of fitting used for joining pipes – capillary and compression. Capillary joints are much cheaper and quite easy to make, but they do require heating with a blowtorch to melt the solder. If you are using a blowtorch, you must take special care, particularly in lofts and under floors, to avoid the risk

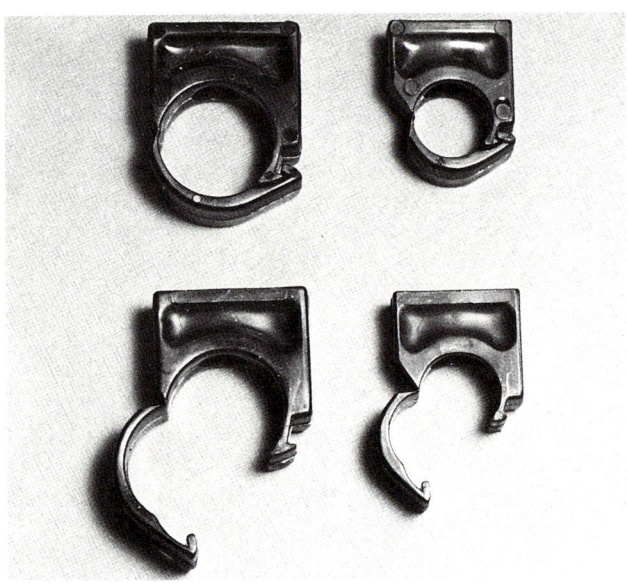

When you are running pipe underneath floors, you should carry it along channels, clipping the pipe at regular intervals. If you fit a cover over the pipework, you will find access much easier should any maintenance be needed.

of fire. Compression fittings only require the use of two spanners and need no heating.

The method you use to join lengths of pipe using capillary fittings is explained in the pictures.

Sometimes you may only need to make one joint on a fitting, for example when you are planning to add to the joint after installation. In this case insert short pieces of untreated pipe into the other joints and wrap wet rags around them while you make the joint you need to prevent the rest of the solder melting as well.

One of the problems of making joints in situ is that the heat necessary to do the work could damage neaby surfaces, such as walls or floors, and create a fire risk, as already mentioned. When soldering joints in this

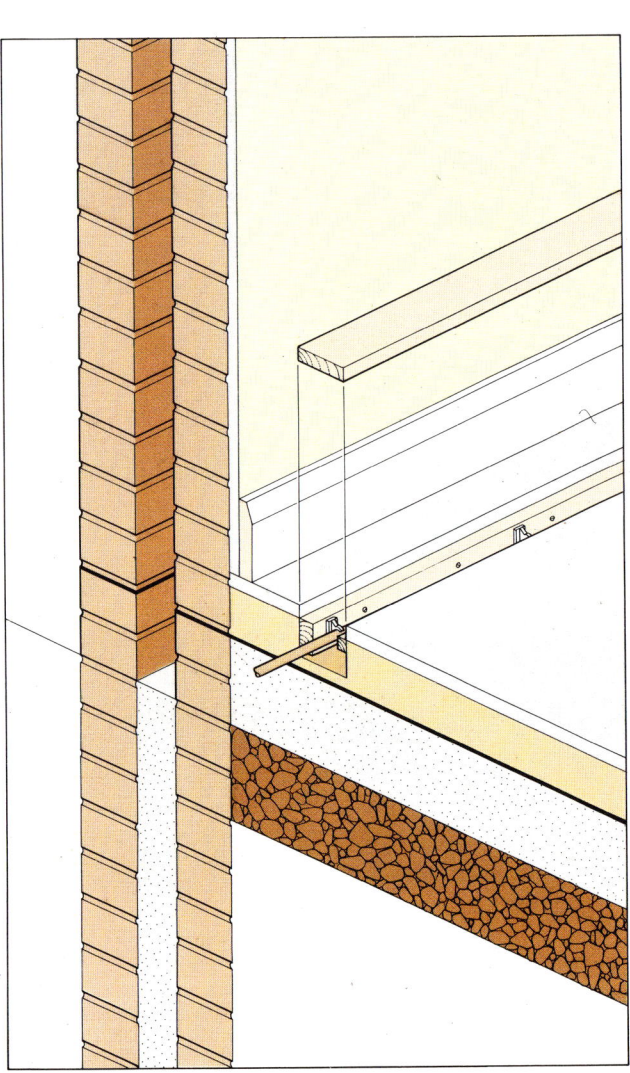

You simply screw these plastic clips to the wall or solid surface, insert the pipe and lock it in. The clips shown here are designed to accept 15 and 22 mm (or $\frac{1}{2}$ and $\frac{3}{4}$ in) pipe.

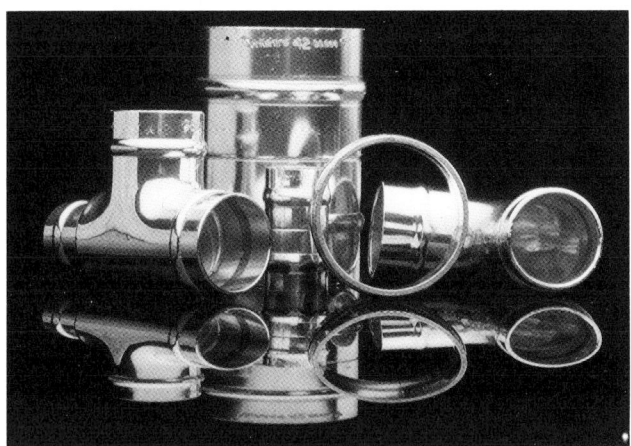

These capillary fittings work on the principle of integral solder flowing around the joint as heat is applied to the join.

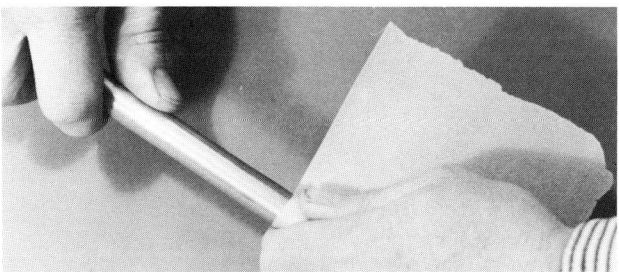

When making a capillary joint, first clean up the cut pipe end with glasspaper or steel wool.

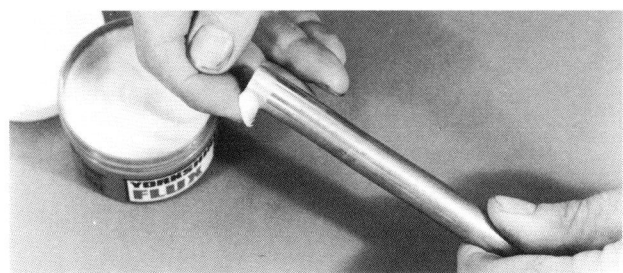

Apply the flux to the outside of the pipe and the inside of the fitting.

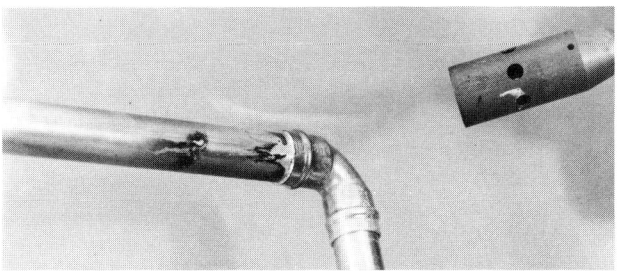

Heat both the capillary fitting and the pipe itself until a complete ring of solder appears at the join. Allow the new capillary joint to cool before cleaning it up with glasspaper or steel wool.

situation, you must use an insulating mat behind the joint as protection against fire.

If you find a capillary joint starts to leak or you have not formed a complete ring of solder at the mouth of the fitting, you can add more solder to effect the necessary repair. But the area round the joint must be dry, clean and fluxed or the solder will not flow into the joint.

Compression fittings are available in brass or other metals where required in those areas in which brass is attacked by the water. If the water is soft, you should first check with your local water authority as to whether you will need to use special gunmetal or dezincified fittings instead.

Each fitting incorporates a socket with a pipe stop to ensure the correct amount of pipe fits into it. A compression ring (sometimes known as an olive), which has tapered ends, fits closely over the pipe and is then squeezed by the nut to ensure it grips the pipe firmly.

If you make this type of joint correctly (see

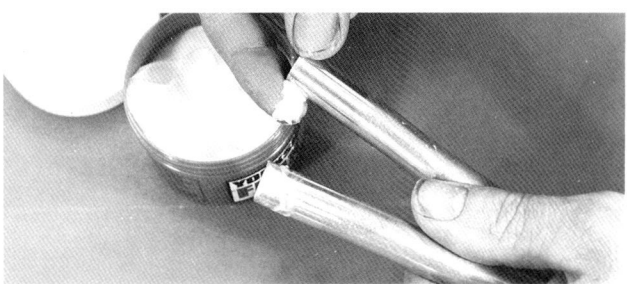

When making an end-feed joint, apply flux to the cleaned end of pipe and inside the fitting.

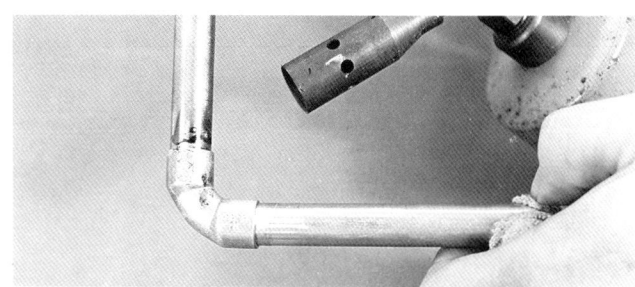

Apply heat to the pipe and fitting until the flux starts to melt.

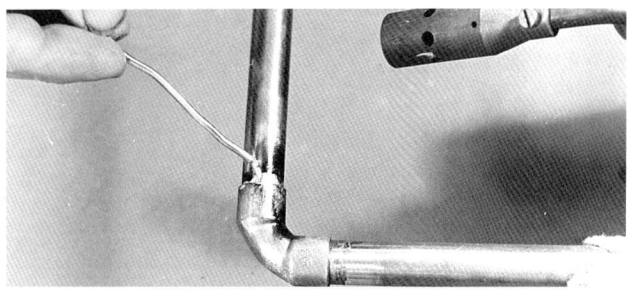

Remove the heat and apply the solder. If it does not flow, remove it and reheat the joint.

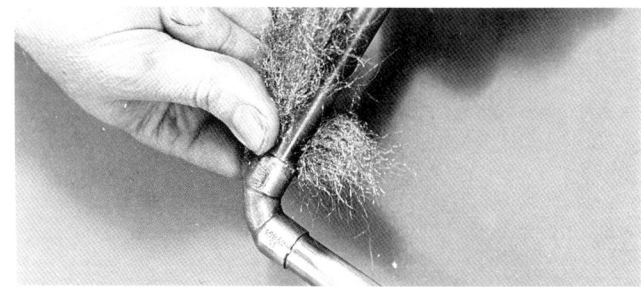

Apply the solder until a complete ring appears round the fitting. Clean the joint when cool.

pictures), you should not need to use any jointing compound.

To guarantee a well-made joint, you must make sure that the pipe is completely round and not damaged or dented in any way, that it is inserted right up to the pipe stop in the fitting and that the component parts of the fitting are square to the pipe. This means the nut should rotate freely before gripping the compression ring. Never overtighten the nut, although with stainless steel pipe it will have to be tightened more than with copper pipe.

One point to remember is that compression rings and nuts are not always interchangeable between fittings from different manufacturers.

Valves

There are several different types of valve used to control a central heating system, all of which perform a specific function within the circuit.

Gate valves   These manually operated valves are used to shut off the flow of water through the pipes and should always be left either fully open or fully closed. The typical locations for the valves are below the cold water storage cistern in the feed pipe to the boiler and hot water cylinder and on either side of the pump, where fitted.

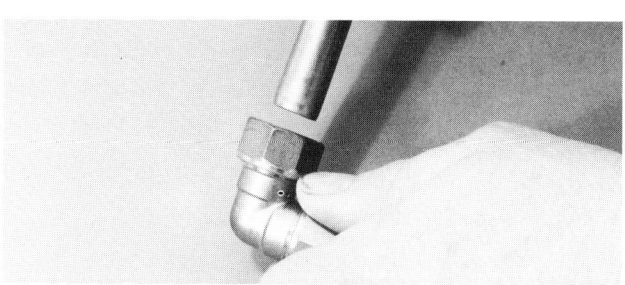

When making a compression joint, insert the pipe into the fitting up to the pipe stop.

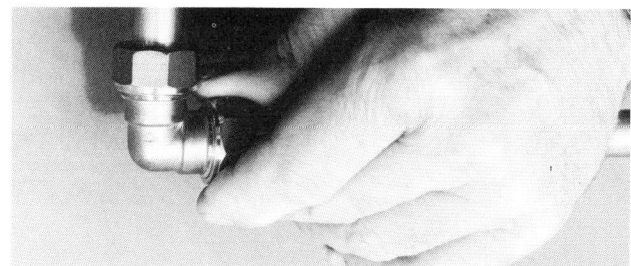

Tighten the nut by hand until the compression ring grips the pipe and stops it rotating.

The range of compression fittings includes those non-manipulative ones made from a dezincification resistant copper alloy (top) and the more traditional brass ones (above).

Hold the fitting with one spanner and tighten the nut with another about two-thirds of a turn. If this join is not completely watertight, you can tighten it further with a spanner.

29

Gate valves control the flow of water and should be left either fully open or fully closed. Normally they are fitted below the storage cistern to isolate it and either side of a pump should repair or replacement be needed.

The radiator shut-off valve (top) controls the flow of hot water through the radiator – and therefore the temperature given off. The lockshield valve enables you to balance the flow of water through the radiator.

The draincock is a special valve that enables you to drain off the system, should any repair or maintenance work be required. Normally one should be fitted at the lowest point on the central heating system, often near the boiler.

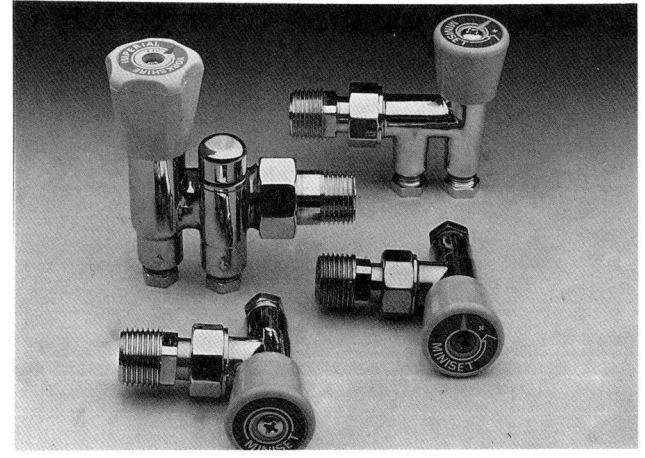

These special radiator valves are for use on systems using microbore pipework. The two at the top are combined flow and return valves. These are fitted to one end of the radiator in a microbore installation.

The air-release valve is fitted to the top of the radiator and is operated by a special key. From time to time air will get trapped in the system. By turning the key you can release the unwanted air through this valve.

Draincock   This valve should be fitted at the lowest point in the central heating system to enable water to be drained off when maintenance or repair work is required. It incorporates a tail on to which you can clip a length of hosepipe and it is opened or closed by means of a spanner.

Radiator shut-off valves   As the name implies, these valves are fitted on to each radiator to control the flow of water through it. This enables you to turn off or isolate individual radiators around the house as required. They are operated by a hand wheel, similar to the handle of a shrouded-head tap but smaller.

Thermostatic radiator valves   These perform a similar function to the shut-off valves, which they would replace. The difference between them is that these have a thermostatic control, which means the flow can be stopped according to the temperature required.

Lockshield valves   These are fitted to the opposite side of each radiator to the shut-off or thermostatic valves and must be set when the central heating system is first commissioned (see pages 108–111). They are then capped and should not be altered unless you need to remove the radiator, in which case they must be closed.

Air-release valves   Also known as bleed valves, these are fitted to the top of each radiator and enable air to be bled out of the circuit and radiator. They are operated by a special key.

There is another type known as an automatic air-release valve, which can be fitted at the highest point in a circuit where pipes drop down to the radiators. These will vent air automatically, but will not allow any water to escape.

Cistern valves   The flow of water into the cold water storage cistern and the feed and expansion tank has to be controlled. This is done with a valve in the cold water supply pipe as it enters either the cistern or the tank.

There are several types of valve available, with the traditional Portsmouth valve being replaced with the up-dated design of the Garston valve. The latter is made of plastic and has the advantage of being easier to service and adjust.

If the pressure in the mains water is particularly high, it would be best to use the Torbeck valve, since this is designed to overcome high pressure and any fluctuation in the water pressure.

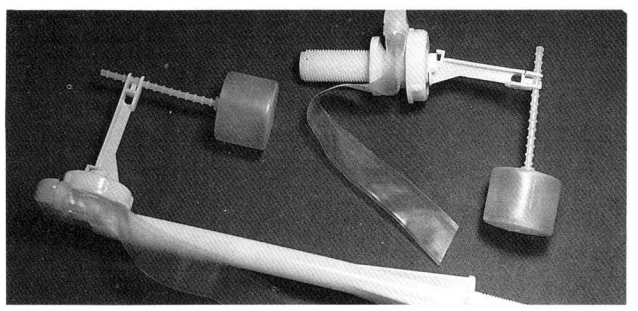

The Torbeck valve is designed to overcome fluctuations in water pressure.

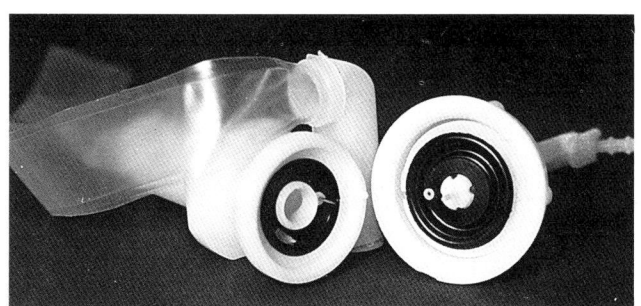

Water pressure acts equally on both sides of the Torbeck valve's diaphragm.

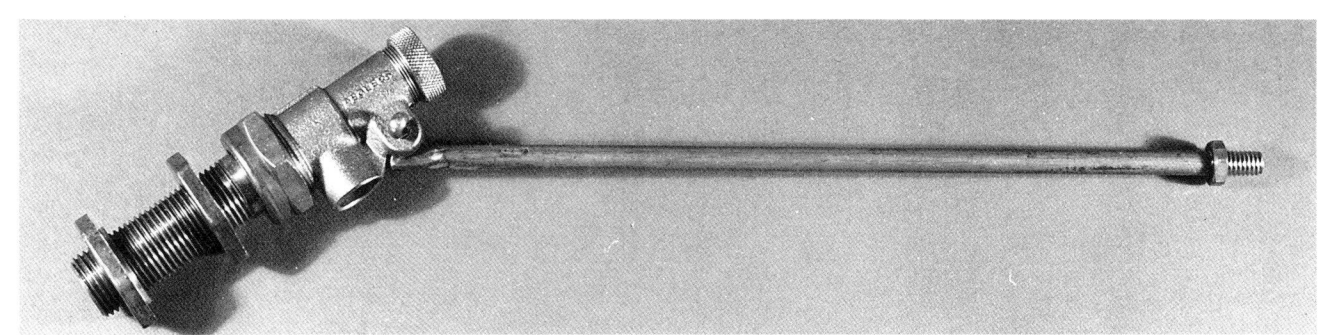

The Portsmouth valve has for years been the traditional means of controlling the flow of water into most cisterns and water storage tanks. Recently, with the development of the Garston and Torbeck valves, it is being superseded. The latter has the advantage of being simpler to adjust. The Portsmouth valve operates by means of a ball float on one end. As the level of the water rises it pushes up the float arm until the valve closes, shutting off the water.

3 Boilers and flues

As already discussed in Chapter 1, you do have the choice of different types of fuel. There are, however, several factors you must consider before making a decision – including cost comparisons. One useful source for this information is the Department of Energy, who produce a booklet called *A Guide to Home Heating Costs*.

Here we look at the main types of boiler available for use with a central heating system, involving four different fuels. If you are in any doubt, however, or need more information on individual fuels, you should seek further advice from the relevant fuel authority.

Solid fuel boilers

Traditionally this fuel was used to fire boilers and it still provides an efficient form of heat. One potential disadvantage was the messy job of loading and clearing out, but nowadays modern designs have made this type of boiler a lot cleaner to operate and maintain.

One form of solid fuel is coal. This is available as Housecoal or its smokeless counterpart, Homefire; these are normally only used on open fires. Another smokeless fuel for open fires is Coalite, although nuts are made for use with room heaters and boilers. Rotal, Sunbrite, Rexco nuts and Rexcobrite are all dual-purpose fuels, the most common being Sunbrite. Phurnacite and Anthracite are produced specifically for room heaters and boilers. When considering a solid fuel boiler, check which type of fuel is recommended for a particular model and whether that fuel is available in your area.

Check on the storage facilities and the accessibility of the fuel store (see pages 7–8) since these are particularly relevant factors to bear in mind. You

should also look at how the ash can be disposed. You can, of course, spread this over the garden. If you are putting it in a dustbin or other suitable container, make sure it will withstand heat. Never put ashes in a plastic bin or waste bag unless they have completely cooled down. Otherwise you run the risk of fire.

Different types of room heater are available for use with solid fuel. The traditional back boiler behind an open fire is capable of heating four radiators as well as the domestic water supply. If you need increased capacity, you can get high output models. The more recent design of room heater has the fire contained behind glass doors, thus offering the joint benefit of radiant heat in the room and the capacity to heat water for both the domestic supply and the central heating.

There are gravity-fed boilers that are designed to fit in with the decor in your kitchen, since they incorporate a stove-enamelled surround. With the latest models, you only need to refuel once a day during the winter and less during the summer. The fuel is kept in a hopper above the small fire, around which is the boiler. There is a damper which controls the rate at which the fire burns.

Solid fuel boilers are not instantly controllable, however, and you will need to provide an outlet for the surplus heat. As previously explained, this can be achieved through a gravity-fed domestic hot water system. One point worth bearing in mind is that when the central heating is switched off, a small heat output is still maintained by the boiler, thus contributing to the overall warmth of the house.

All solid fuel boilers require a flue – either internal or external – which must be kept clean and so must be swept regularly (see Flues, pages 41–45).

The air necessary for combustion is taken from the room in which the boiler is situated. It is therefore important to ensure that adequate ventilation is available in that room. This can be arranged via air vents or ducts connected to the outside of the house and terminating near the boiler or heater. Not only

Housecoal

Anthracite

Welsh Dry Seam Coal

Homefire

34

Phurnacite

Coalite

Royal Superbrite (Rexco)

Sunbrite

35

does this arrangement reduce draughts, but it also means your expensively heated air is not used to fuel the fire.

If a sufficient supply of air is not available, the boiler or heater will not work properly and could lead to problems from fumes in the room where the boiler is sited.

There is a scheme available whereby you can have an open fire back boiler connected into a system fuelled by oil or gas. When in operation, the back boiler saves on oil or gas in the main boiler and still offers the benefits of an open fire in the lounge. If you are interested in the link-up system, you should contact your local Solid Fuel Advisory Service.

Oil-fired boilers

Despite the increase in the price of oil, this fuel is still widely used in central heating systems. Apart from the cost, there are other considerations such as the fact that you will need to have a large oil tank installed – and planning permission may be required for this. Equally the tank must be reasonably accessible for deliveries of oil. You also need to have the oil filter situated outside the house in a situation where it can be cleaned periodically.

With this type of boiler a pump forces the finely atomised oil into a combustion chamber where it is fired by a pressure jet burner with the help of air. Because of the various working parts in the system and the high-powered flame, particularly when it ignites, this type of boiler is not silent in operation.

For this reason, the boiler is normally sited away from living areas in the home. In most cases an external – or internal lined – flue must be incorporated which reaches to above roof level. In some cases, however, manufacturers do produce a low-level flue; this passes through the wall behind the boiler. Planning permission may be required before you can

This oil-fired boiler has been designed to fit neatly into a kitchen or other suitable indoor area. However some people prefer to install this type of boiler outside the living area – sometimes in a convenient outhouse.

install this type of flue and you should check first with your local authority.

The oil-fired boiler needs regular servicing on an annual or twice-annual basis. This should be carried out by a suitably qualified heating engineer. Most manufacturers operate a regular servicing scheme.

If you are considering having an oil-fired central heating system installed, the major oil companies will give you advice on the best type of system and boiler – and offer guidelines for the installation of the oil storage tank.

LPG-fired boilers

Liquefied petroleum gas (LPG) is available as a fuel for central heating systems where the conventional supply of piped gas is not available. If you are considering installing this type of system, it is worth just checking with your local Gas Board to make sure it is not planning to run a gas main through your area in the near future, since this would save considerable expense.

The gas normally supplied is propane, which is available in cylinders that can be exchanged when empty. You can have your own cylinder installed – at some considerable expense – and get it refilled when necessary. The positioning of permanent cylinders and exchange ones as well is subject to stringent safety guidelines. Always contact your local supplier first, before going ahead with the installation of this type of system.

The boilers available are similar to the range of standard gas boilers, but you should bear in mind that they are not interchangeable. You can get either floor or wall-mounted models, which are available with open or balanced flues (see below), and some incorporate an economy device that ensures that the small pilot flame only burns when the main burner is required, thus saving on fuel.

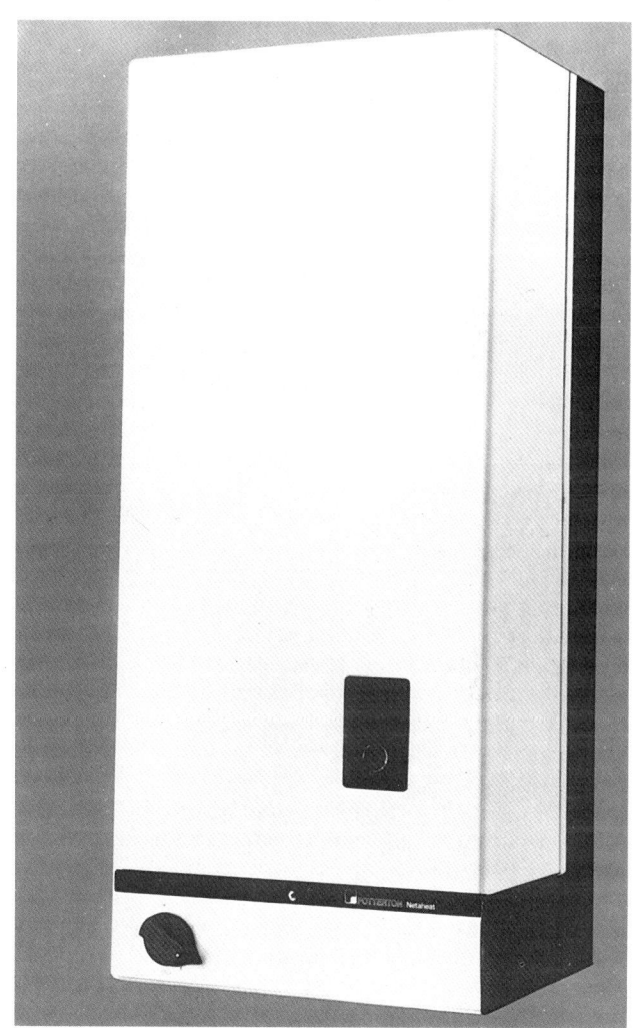

LPG-fired boilers look very similar to standard gas models, but they are not interchangeable. Relatively expensive to run, they could be one answer where mains gas is not available.

Gas-fired boilers

Gas is the most common fuel used for central heating systems. It is clean, virtually silent in operation and instantly controllable. Its use, however, is strictly covered by safety regulations and its installation should be left to a qualified gas engineer. Even if you do install the boiler yourself, the checking of the system and final connections should be made by the local Gas Board. For this reason, the fitting and connection of a gas-fired boiler has not been included in this book.

Gas boilers fall into three main categories: back boilers behind gas fires fitted in conventional fireplaces, wall-mounted boilers and free-standing models.

Gas fire with back boiler This type of boiler fits into an existing fireplace. The gas fire is mounted on the front and the two can be controlled independently – that is, you can have either the fire or the boiler on, or both.

When it comes to the installation, you will have to remove the old firebricks and provide a sound level base for the boiler. The installation drawings provided with the model of boiler you buy will indicate the size of the necessary openings for the heating pipes.

Make sure you have the chimney thoroughly swept and, if it is not already salt-glazed, then you will need to have it lined. The flue must have an approved capping design.

The air necessary for combustion will be taken from the room in which the boiler is installed and you must ensure that there is an air vent fitted nearby with a minimum area as specified by the boiler manufacturer. This vent can be linked directly to the outside or pass into an adjoining room that is linked to the outside. In the latter case, check any ventilation arrangement with your local Gas Board first.

All the modern gas boilers have flame-sensing

This gas fire, which fits into an existing fireplace, will heat the room with radiant heat and by convection, while the back boiler inside the fireplace provides hot water for the domestic and central heating systems.

devices which automatically cut off the gas supply to the burner if there is no flame. This cut-out system will also operate in the event of an electrical failure. With some boilers the pilot light is only lit when the main burner is required. Because this burner can only operate when the pilot light is on, this not only acts as an additional safety device but also saves gas.

Wall-mounted boilers   As the name implies, these smaller boilers are hung on the wall and they are available in a range of heights to suit individual locations – such as under a work surface or in a cupboard.

Because of their low water content, some boilers are only suitable with fully pumped systems, although other models are available for use with gravity-fed systems. In such a situation, however, you will need bypass pipework to ensure continued water circulation after the domestic and central heating systems have been shut down (see pages 108–111).

Most wall-mounted boilers are fitted with balanced flues and are sometimes known as room sealed boilers. These have to be hung on an external wall, with the two-channel flue connection passing horizontally out of the back – or side – of the boiler through the wall. Air for combustion is drawn in through one channel and the products of the combustion are emitted through the other channel. You can incorporate a fan to assist in the emission of waste products. The joy of this arrangement is that no air for combustion is taken from the room in which the boiler is sited.

Balanced flues can be installed in most locations, but here are a few points you must consider before planning where to fit one:
- The flue must be at least 300 mm below a window or air inlet.
- The flue must not be fitted beneath the eaves of a balcony or in re-entrant corners of the building.
- The flue must not be lower than 300 mm above ground level or closer than 600 mm to any corner.

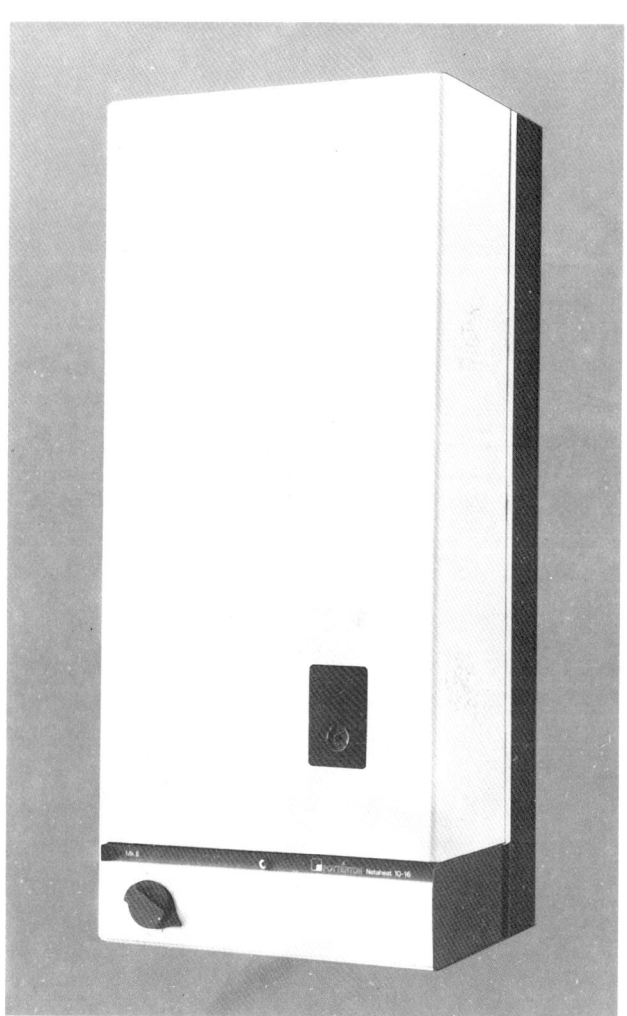

Wall-mounted boilers have the obvious advantage of saving space and can therefore be sited in relatively confined areas where a free-standing model would not fit. This gas design can normally be fitted to any outside wall.

- The flue must not be within 600 mm of a facing surface, for example in a narrow passageway.
- If the flue is within reach of passers-by, it should be fitted with a special guard to prevent people burning themselves on the hot metal.

You can, of course, overcome these potential problems by installing a wall-mounted boiler with an open flue.

Most of these boilers can be fitted between or inside hanging cupboards, which means that the pipework can be concealed. Before you buy a particular model, however, you must make sure that it is suitable for the system you want to install.

Free-standing boilers   These operate on a similar basis to wall-mounted ones and are available with open or balanced flues. Again, some boilers may need bypass pipework to ensure continuous circulation through the boiler.

The boilers are normally supplied in casings of a suitable size to blend in with standard kitchen units and work surfaces and are designed to stand on a flat, non-combustible floor surface. If you have a wooden floor, you must check on the instructions given in the installation leaflet. This is supplied with all makes of boiler and gives detailed and comprehensive advice, so you must read it thoroughly before starting any work on fitting the boiler.

If you are not incorporating a balanced flue, you must check carefully on what requirements are necessary for ventilation.

Isolated boilers   You may decide, because of lack of space in the house or because of the noise level while it is operating, to install the boiler in a separate area such as a boiler house. If this is the case, there are certain factors you should bear in mind.

The area in which the boiler is to be installed must have a non-combustible lining with a half-hour resistance to internal fire. This lining should be fitted

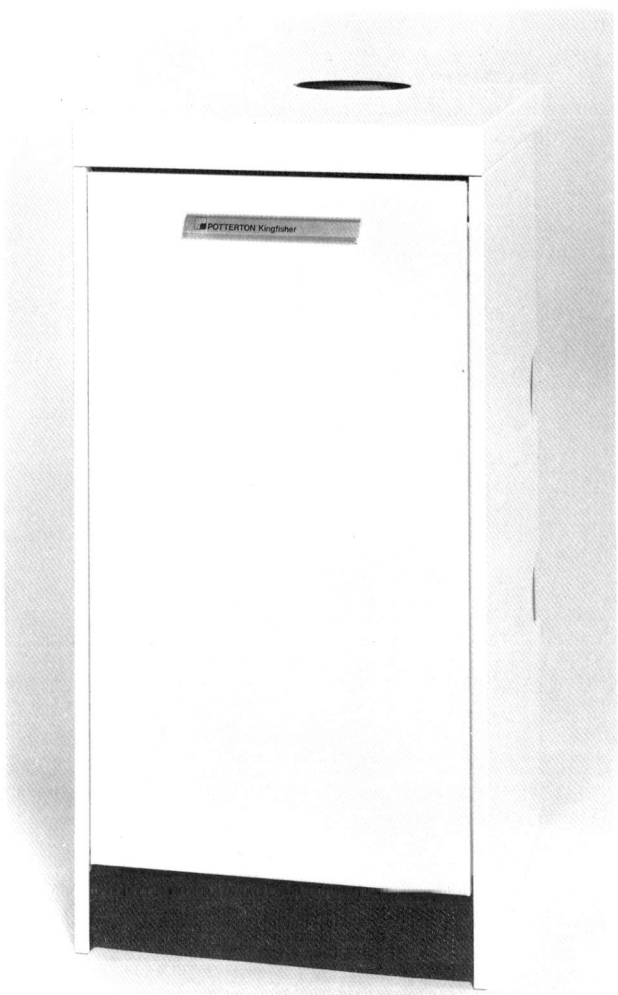

This free-standing gas boiler is designed to fit neatly into most kitchen layouts, ideally between or at the end of work surfaces. The range of models is available with either open or balanced flues.

to the door as well which, incidentally, must be large enough to allow the boiler to be carried through. Ideally the place should be built of brick or similar material, plastered internally and with a flat solid floor. You should make sure there is sufficient access around the boiler to enable servicing to be carried out easily.

You must also adhere strictly to all ventilation requirements. If you fit a boiler with a balanced flue, then arrangements must be made to provide enough ventilation to allow the heat from the boiler to escape.

One point you should not forget if you house a boiler outside or in an unheated area away from the living part of the house is that it must be sufficiently well insulated to prevent possible damage from frost in the cold weather (see pages 8–9). You should also fit a frost thermostat to switch the boiler on when the outside temperature drops below freezing point.

Flues

If you cannot install a balanced flue, you will have to make alternative arrangements to dispose of the waste products from the boiler. Ideally these should be contained within the fabric of the house, since this will reduce the level of condensation inside the flue. The condensation is caused by the excessive cooling of the flue gases, which contain a proportion of water vapour.

The simplest way of achieving this is by using an existing chimney or flue pipe. This must, however, be protected with a suitable flue lining, unless it is salt-glazed and in a good state of repair. The instructional leaflet with the boiler will give advice on the recommended sizes of flue lining. You will have to measure the length of the chimney to determine the amount of lining you require.

This is most easily done by lowering a weighted piece of cord down from the top. To do this, you will probably have to remove the chimney pot and flaunching. Be careful when you do this, since chimney pots are heavier than they look.

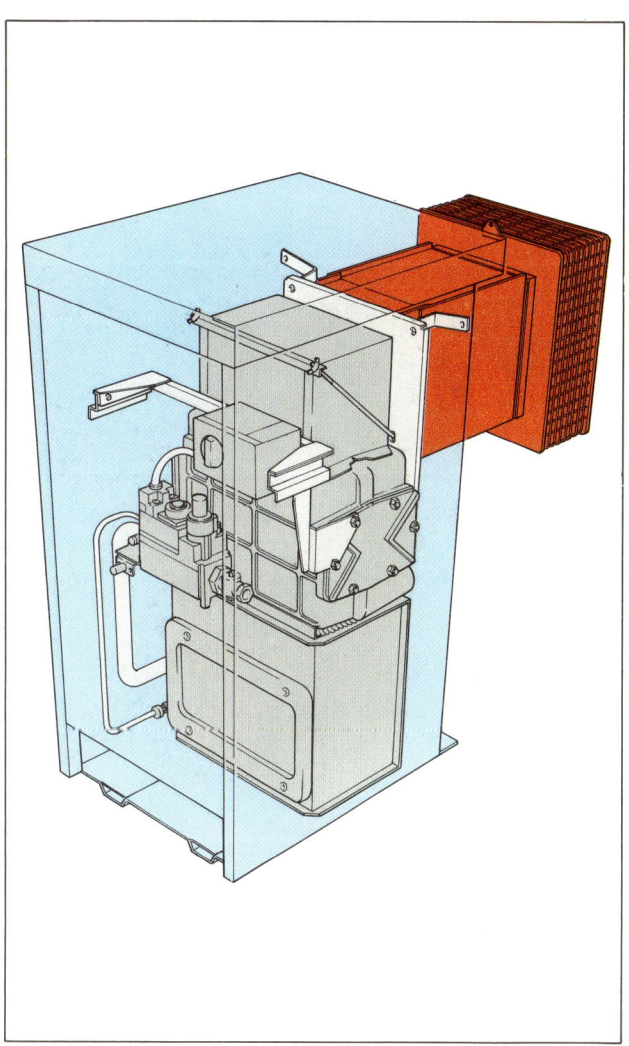

The advantage of a balanced flue gas fire or boiler is that the air for combustion is drawn from the outside and the waste products emitted back again. The flue, which must pass through an external wall, is shaded red.

41

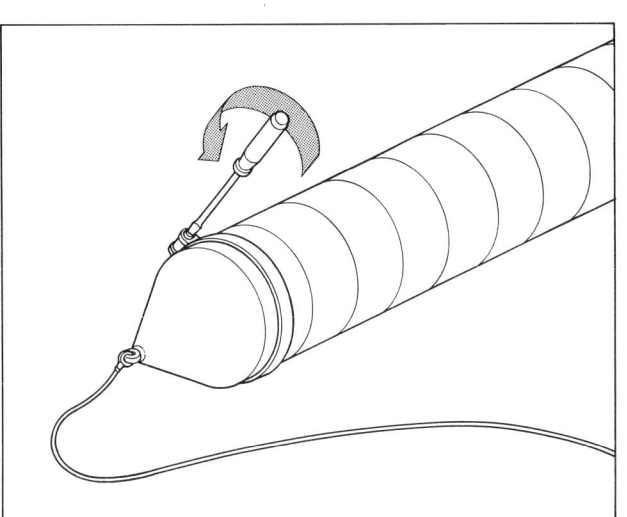

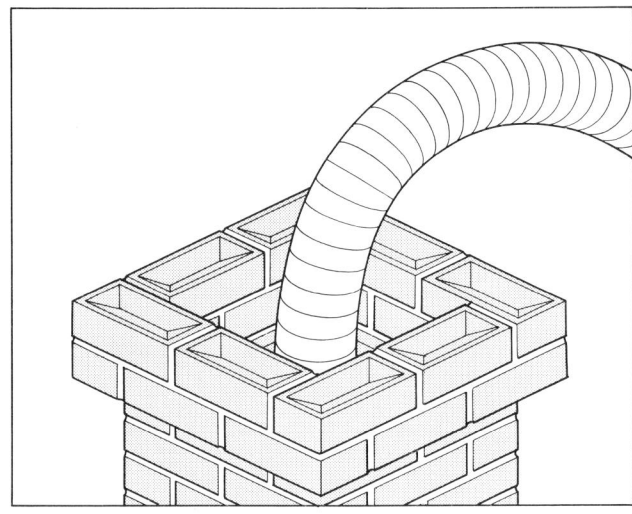

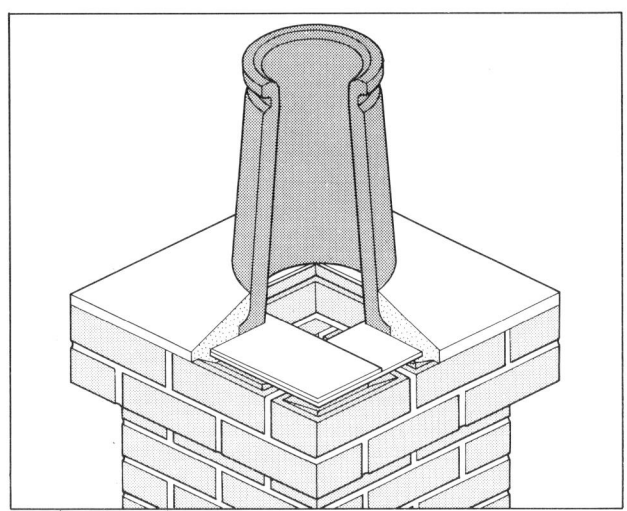

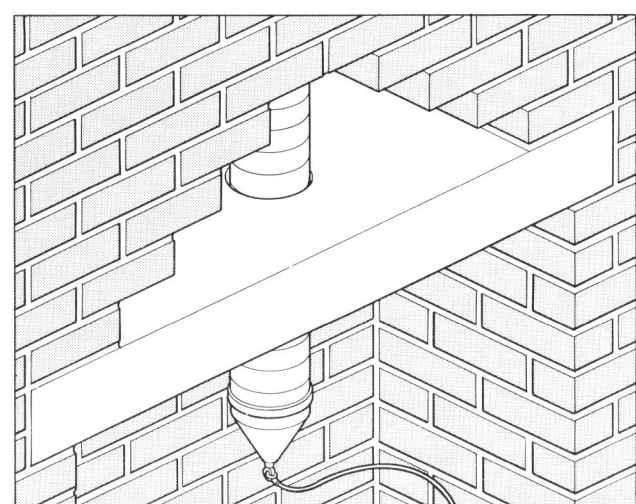

When installing a flue lining in an existing chimney, you first have to fit the dummy nose cone into the end of the liner (top left), tightening the retaining clip to hold it in place. Then carefully remove the old chimney pot by chipping away the flaunching (bottom left). Lower the cord down the chimney and feed the liner through (top right) by pulling the cord and pushing carefully from above until it is in position in the fireplace (bottom right).

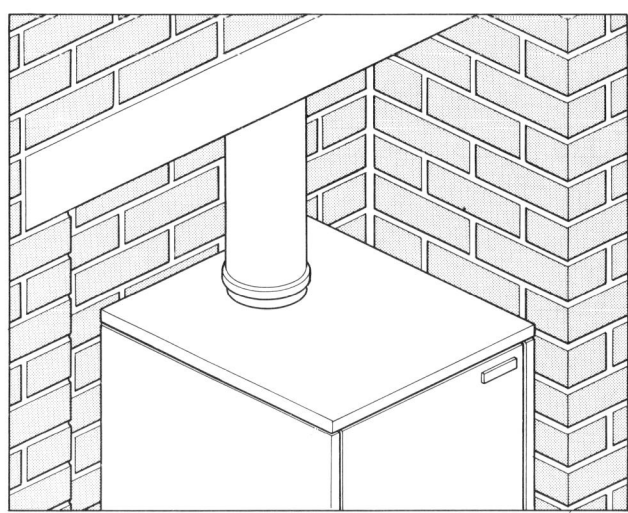

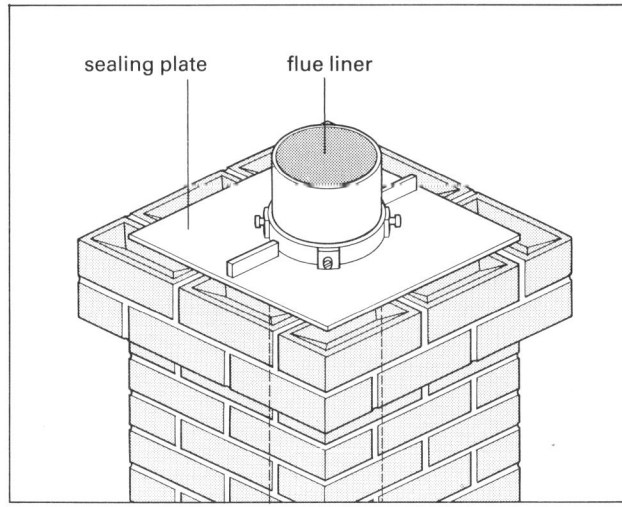

sealing plate   flue liner

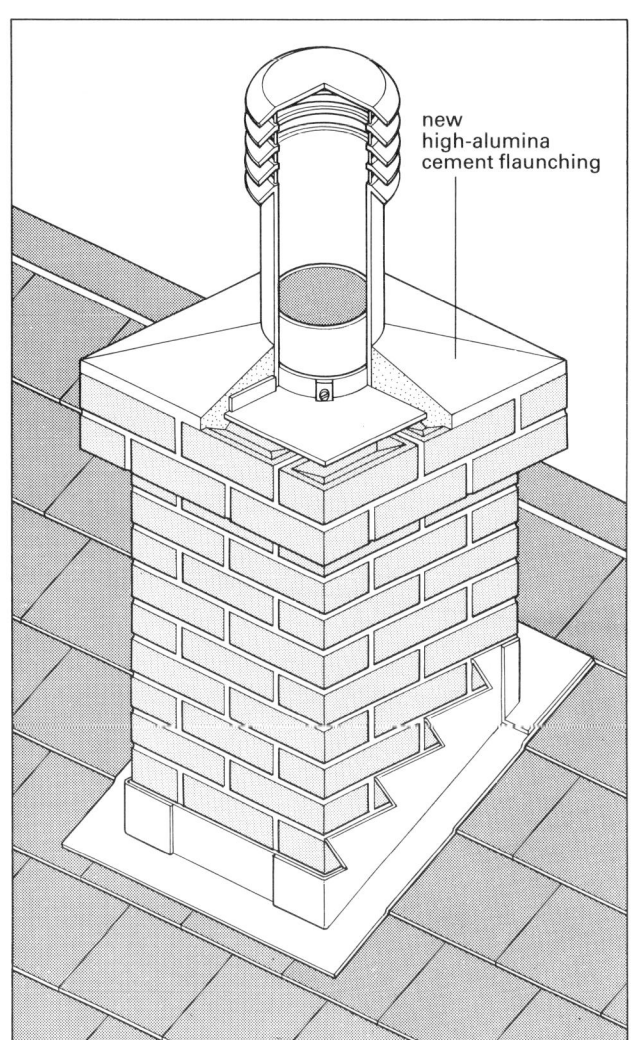

new high-alumina cement flaunching

Secure the liner in position in the fireplace using the correct connections at the top of the boiler and the top of the chimney stack (above). The liner has to be held in place at the top of the chimney with a special terminal

and this fitted to secure the liner with high-alumina cement (right). Since you are working at considerable heights, great care must be taken with regard to safety. You will need somebody outside to relay messages.

When putting in a flue lining, carefully fit the wood or plastic nose cone into the bottom end of the liner to prevent it collapsing as it is drawn down the chimney. Then lower the pull cord down the chimney.

At this stage in the operation you really need three people to complete the installation. One should be at the top of the chimney feeding the liner down it; one should be inside gently pulling the liner down the chimney with the cord, and the third should be outside the ground floor room relaying instructions.

When you have got the liner in place, tie the top in position while you fit the sealing plate. You can then remove the pull cord from the bottom of the liner and secure the sealing plate and correct flue terminal into position with new flaunching, using high-alumina cement. Then cut the bottom of the liner as required and fit it to the special boiler connection.

If there is no suitable flue available, you will have to install an external one. Never attempt to make use of an old or existing single thickness asbestos flue. Where you need to fit a new flue, check first with the local authority since you will probably need planning permission for it. You should also seek advice from the boiler supplier and your local builder, since many variations are possible. The external flue pipe must be insulated and it may also be necessary to provide a condensation drain to remove excess moisture. Bear in mind, too, that recent regulations on the use of asbestos may limit your choice of materials.

The outlet from the flue must be above the level of the roof and at least 600 mm (2 ft) above the intersection of the flue and the roof. With a flat roof, the flue must terminate at least 250 mm (10 in) above the roof surface – or 600 mm (2 ft) if the roof has a parapet.

Where you have a complicated roof structure or the roof is in a particularly exposed situation, the flue terminal must not be fitted in such a way that the wind can create high pressure down the flue. You will find that the type of flue terminal to be used will normally

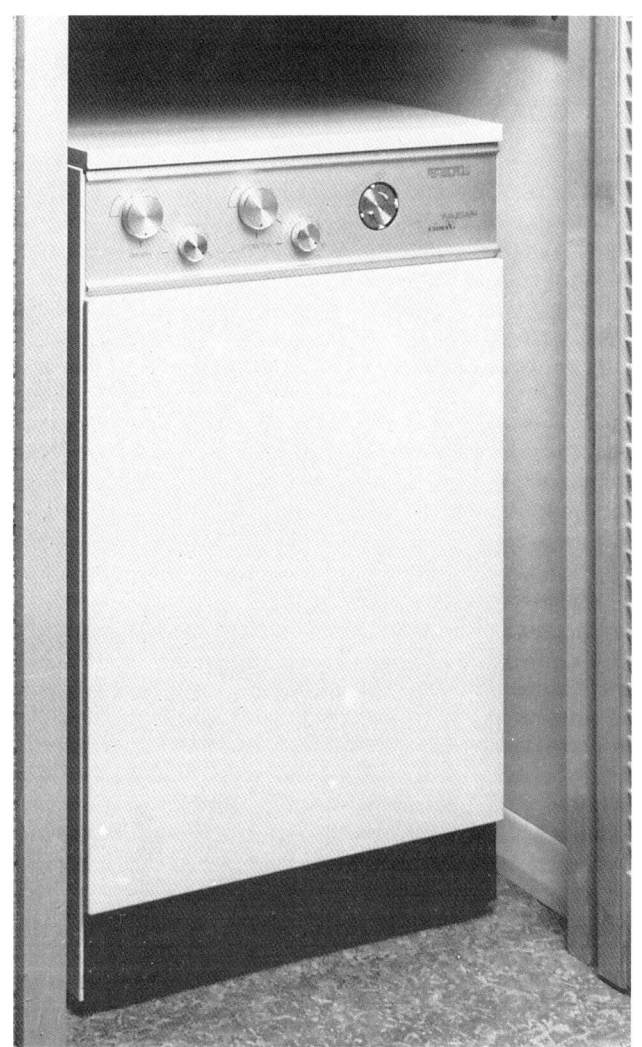

The joy of a gas combination unit is that it is completely self-contained and requires no cold water storage cistern, feed and expansion tank or hot water cylinder. And it will fit neatly into a reasonable size airing cupboard.

be specified by the boiler manufacturer. It must be of a design that allows free discharge of waste products from the boiler, minimises downdraughts and prevents anything getting in through the top.

Gas combination units

Several of these units are now available that are completely self-contained and do away with the conventional hot water cylinder and expansion tank – in some cases the cold water storage cistern as well. The units only ignite when hot water is required for either domestic or central heating purposes.

Manufacturers do claim lower installation and running costs for the units. They can often be fitted in an airing cupboard in place of the hot water cylinder, thus saving valuable space in the kitchen, for example.

For further information on the combination units you should contact your local Gas Board, where you can get full details.

4 The controls

Before you can finalise the design of the system, you will have to decide what controls you want to incorporate. The range is extensive and growing – with new systems and developments coming on the market all the time.

As a general rule, the simpler the control system you use, the easier it is to install and run. You will, of course, have to weigh this against the fact that the more sophisticated the system, the more control you will have over such things as the temperature level in the individual rooms and that of the domestic hot water. Also to be considered is the fact that you will want to use the fuel in the most economical way, while ensuring the optimum degree of comfort and efficiency.

Here we look at the basic philosophy of controls, starting with the simplest methods and working up to the more sophisticated types of control. You must, however, check carefully with the boiler instruction leaflet against the control system you want to incorporate to make sure they are compatible. Bear in mind too that most boilers are supplied with their own inbuilt system – and this may duplicate some of the functions you have planned to include.

A basic system

There are two functions of the system that must be controlled above all others – the operating time of the boiler and the temperature of the water. If, in addition to these two items, the central heating pump is controlled as well, then you have your basic control system.

Timing is usually controlled by an electrically operated timeswitch which, as a minimum, gives you two sequences for switching the system on and off in

This basic timeswitch enables you to preset the central heating system to switch itself on twice in every 24 hours at predetermined times, when fitted to a simple gas boiler arrangement.

every 24-hour period. The boiler thermostat controls the temperature of the water, turning the boiler off when the desired water temperature has been reached and on again when it drops below that level.

The central heating pump can be controlled by a simple on-off switch wired from the main timeswitch. With this arrangement, when the pump is switched on it will run continually all the time the boiler is switched on, thereby pumping hot water round the system through the radiators.

You can switch the pump off during the summer months, when the central heating is not required, and the boiler will heat up just the domestic hot water as needed.

One refinement to this system is the installation of a gate valve in the upstairs central heating circuit. With this control you can heat up the ground floor, for example, without wasting fuel by heating the bedrooms as well. This can be of particular value during the daytime and in the autumn, when the bedrooms do not necessarily need heating.

The control system described here should only be used in conjunction with skirting or convector radiators, since the water in the primary circuit may have to be above 80°C at times, in order to give sufficient heat input to the domestic hot water calorifier. Such a system would work adequately in a small, well-insulated house (particularly if you only require background central heating), but it does have the following limitations:

- no control over the temperature of the domestic hot water
- no control over basic room temperatures
- no variation in individual room temperatures once the installation has been completed
- the timeswitch has to be adjusted manually if at certain times you want different heating periods
- the basic on/off control for the central heating has to be operated manually.

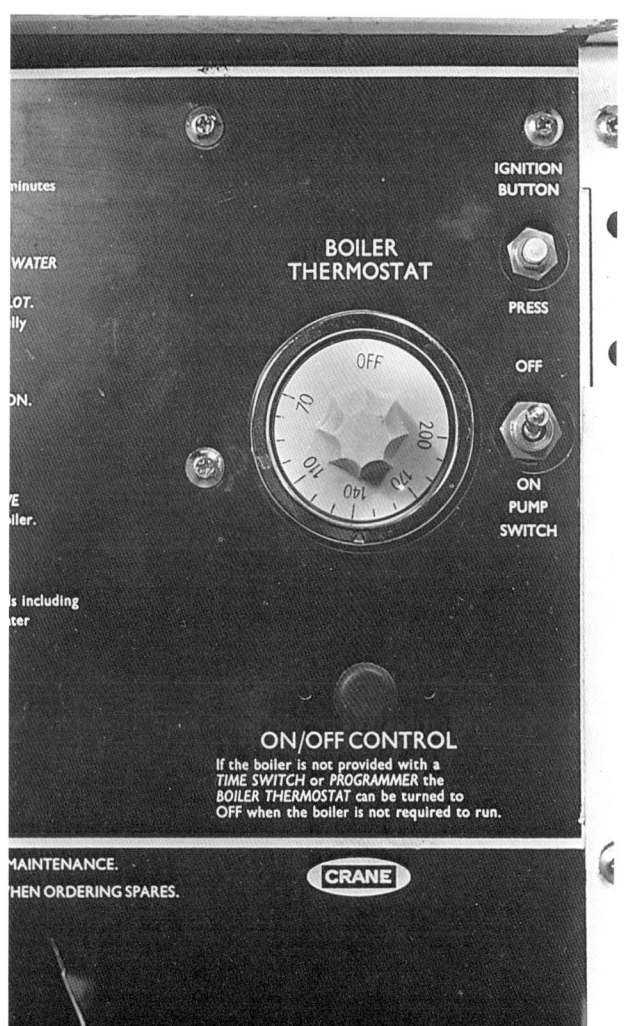

The boiler thermostat controls the temperature to which the water in the boiler is heated. This particular model, which is fitted to a basic gas-fired central heating system, can be used with or without a programmer/timeswitch.

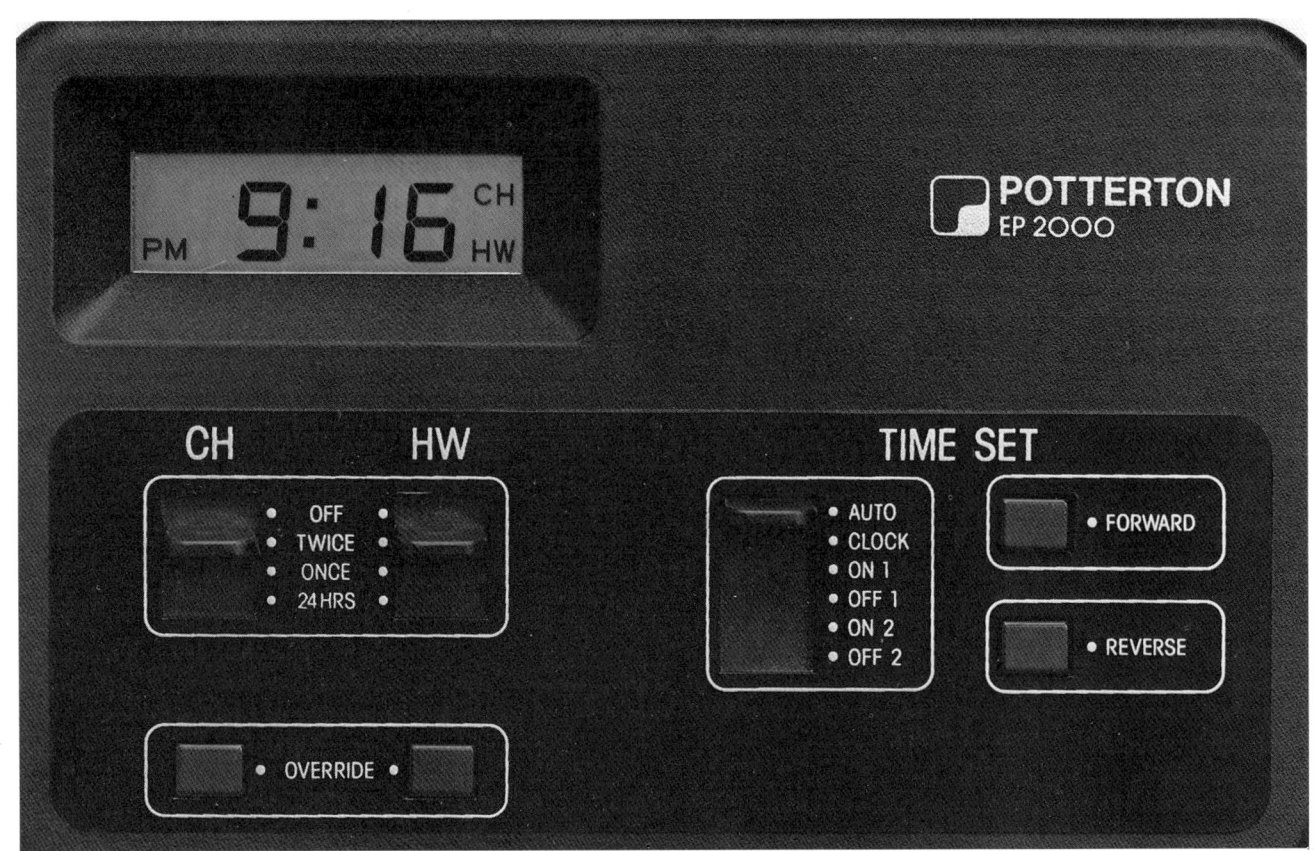

The latest programme controllers are electronically operated to enable you to alter the running of your domestic hot water and central heating systems. The great *advantage of these latest models is that they offer a whole range of programmes.*

All of these can be overcome with the use of more sophisticated controls, which are outlined below. The method of connection, including wiring, does vary and you should check the instructions given with each type of control before attempting installation. For this reason, no fitting or operating details have been included here.

Timeswitches and programmers

These all incorporate a clock – and in most cases this is a normal 24-hour electric clock, although digital ones are also available. Included in the clock are devices to switch the boiler on and off and these can be set for any time – day or night – during each 24-hour

Programmes available on 10–16 programme controller

Programme	Hot water	Central heating
1	OFF	OFF
2	ONCE	OFF
3	TWICE	OFF
4	CONSTANT	OFF
5	ONCE	ONCE
6	ONCE	TWICE
7	TWICE	TWICE
8	CONSTANT	ONCE
9	CONSTANT	TWICE
10	CONSTANT	CONSTANT

Additional programmes available

11	OFF	ONCE
12	OFF	TWICE
13	OFF	CONSTANT
14	ONCE	CONSTANT
15	TWICE	ONCE
16	TWICE	CONSTANT

period. They control two separate circuits – one for domestic hot water and one for the central heating. By programming the control, you can vary the set times at which the heating and hot water are switched on and off. The diagram shows typical programmes available on a 10 and 16-programme controller.

These programmes are suitable for all gas and oil-fired boilers. In the case of solid fuel boilers, the domestic hot water will be available all the time the boiler is running, since the heat cannot be turned off.

Thermostats

Most makes of boiler incorporate their own therm stat, but this will only control the temperature of the water in the boiler itself. This means that you will need separate thermostats to control the temperature of the domestic hot water and the central heating circuits. These can be linked to and operate either motorised valves in the circuit or the central heating pump itself.

Cylinder thermostat   This is fitted to the outside of the hot water cylinder and held in place by a steel band which passes right round the cylinder. It is either fitted beneath the insulation jacket or incorporated into the covering of a pre-insulated cylinder – depending on the type. Alternatively it can be fitted to the return pipe from the calorifier to the boiler. When the water in the cylinder – or the return pipe – reaches the desired temperature, this thermostat operates a motorised valve to stop the flow of water through the calorifier.

Room thermostat   This has an on/off facility built into the control which is activated when the pre-set temperature level is reached in whichever room the thermostat is situated. The siting of this type of

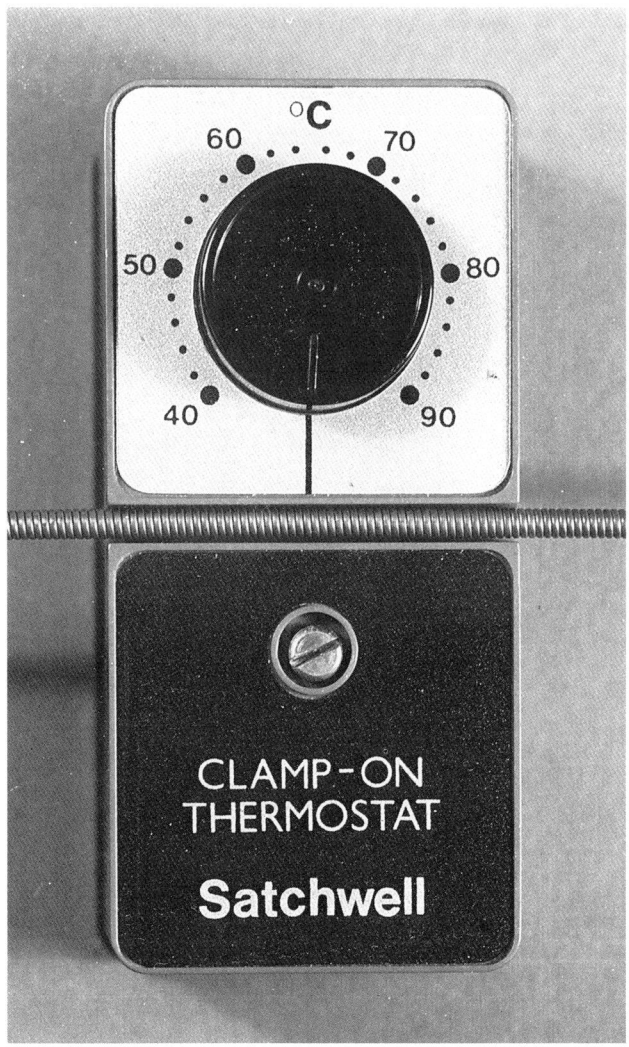

The hot water cylinder thermostat, which with the help of a steel spring band clips round the outside of the cylinder, controls the temperature in the cylinder through a motorised valve.

control is therefore critical and you should bear in mind the following:

- The room in which it is fitted must be heated solely by central heating, since other sources of heat could have the effect of switching off the central heating, leaving the rest of the house unheated.
- The thermostat should be positioned about 1.5 m (5 ft) up from the floor and away from draughts, radiators, wall lights or other forms of heat.
- Normally a maximum of one room thermostat should be fitted for each circuit, since it operates a motorised valve which stops the flow of water through the circuit.
- It can have a variable setting with adjustable stops to determine the highest and lowest optimum temperature required.

Special models are available that incorporate a clock and reduced temperature facility. This enables the thermostat to control the boiler and provide the desired heat levels at particular times through the day – for example, morning and evening. At other times the heating can operate at lower levels, ensuring that an overall minimum temperature is maintained throughout a 24-hour period, if required. This can help in a more economical use of fuel.

If you do install this type of room thermostat, it can make the central heating programmer redundant – or you can switch to continuous heating.

Another type, which can prove useful during the winter months, is designed to protect the system from frost damage. It functions by overriding a system that is otherwise switched off by activating the boiler and central heating when the temperature inside the house falls to 1°C. It is probably only necessary in a situation where the house is likely to be left unoccupied for several days at a time or where it is in a particularly exposed situation. The thermostat can either be incorporated into the standard room type or fitted as a separate control.

This basic room thermostat should be fitted to the wall of a centrally heated room. When the temperature in that area reaches the preset level, the thermostat cuts out the circuit until the temperature drops below that level again.

Thermostatic radiator valves

The thermostatic radiator valve (TRV) is probably the most versatile type of temperature control since it enables you to control virtually every room in the house independently. However, to ensure a continuous flow of water through the circuit, at least one radiator in the system should be left without a TRV.

This type of thermostat is fitted directly to the radiator and replaces the normal shut-off valve. Remember that a lockshield valve must still be fitted on the other end of the radiator. The TRV reduces or shuts off completely the flow of water through a radiator when the room reaches the required temperature. The thermostat itself is contained in the

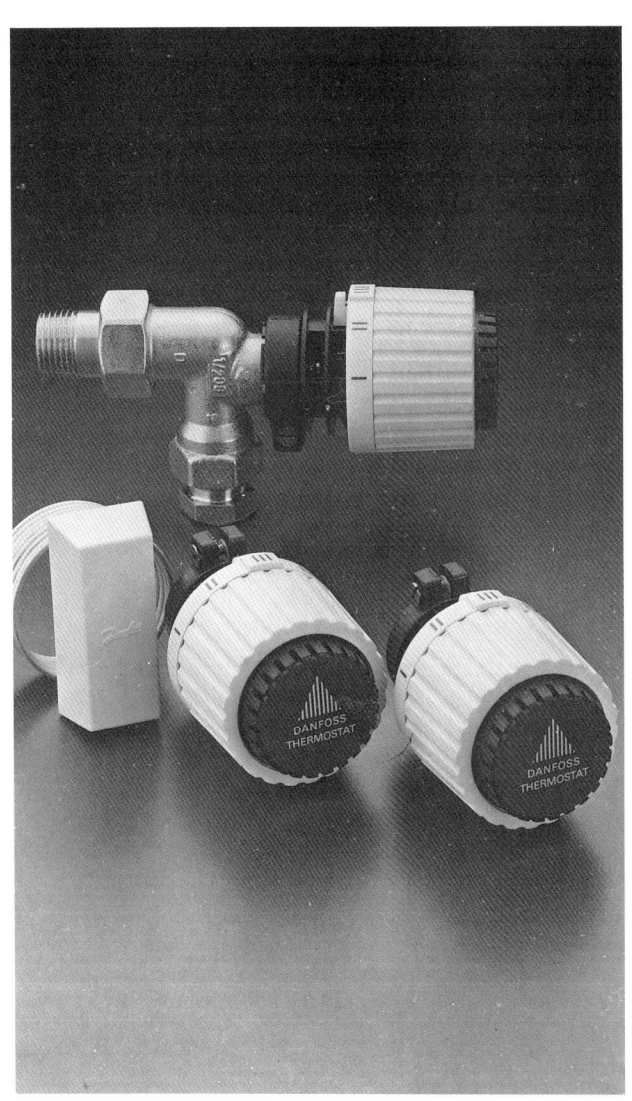

The main advantage of fitting a thermostatically controlled radiator valve is that individual rooms can be kept to a predetermined level of temperature.

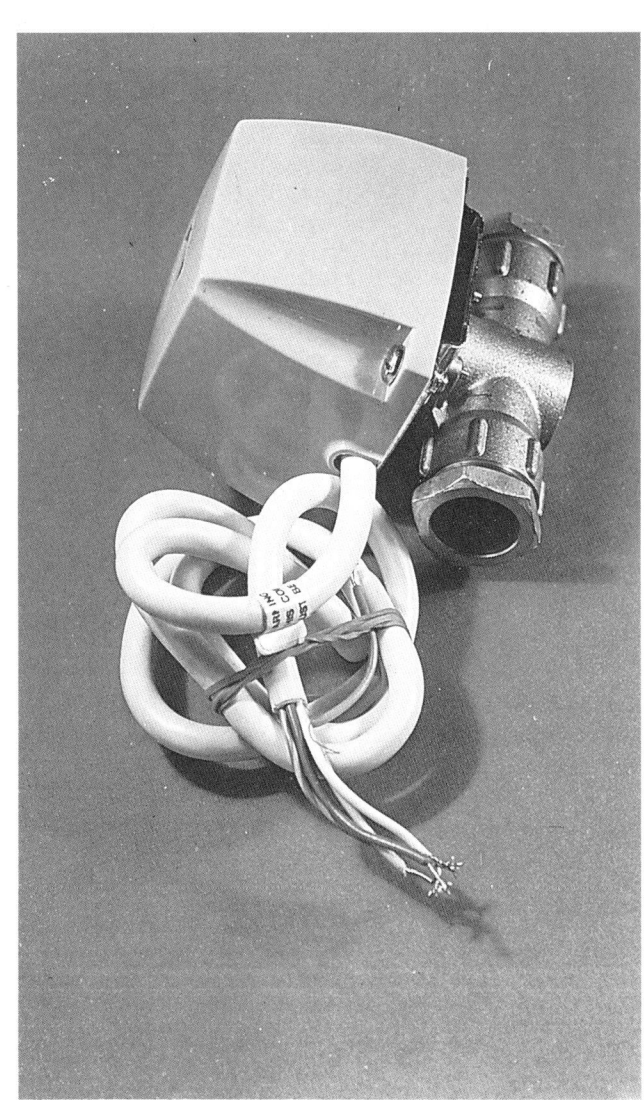

With a remote-reading thermostatic radiator valve, you can achieve a more accurate sensing of the overall temperature in a room, rather than around the radiator.

head of the valve and reacts to the temperature of the room – not the water.

In situations where the TVR might react to misleading temperatures, such as in a corner, a remote-reading TRV is available. With this type, a remote sensor is connected to the valve by a steel capillary tube and placed in a more open position up to 2 m (6½ ft) from the radiator.

Motorised valves

As the name implies, these valves are electrically operated and normally used in conjunction with a thermostat to control a circuit. Three basic types are available.

The simplest valve is fitted into a single pipe and either prevents or permits the flow of water, depending on whether it is closed or open. The diverter valve is normally used in systems where either the domestic hot water or central heating circuit is fed by the boiler. The three-way valve is similar to the diverter but has an added facility. With this model, the flow from the boiler up the stem of the 'T' can be diverted either left or right and can be made to flow in either direction (see diagram). It can be used to control a fully pumped circuit giving either domestic hot water or central heating – or both.

When a three-way valve is fitted with a fully pumped system and both circuits reach their predetermined temperature, the pump and boiler are shut down automatically by the thermostats. If only one of the circuits is at the required temperature, the valve will operate to divert the flow of water through the other circuit.

The same facility can be achieved by fitting two simple valves – one in each arm of the 'T'. But the advantage of the three-way valve is that it is impossible for the pump to work against a closed circuit. This is

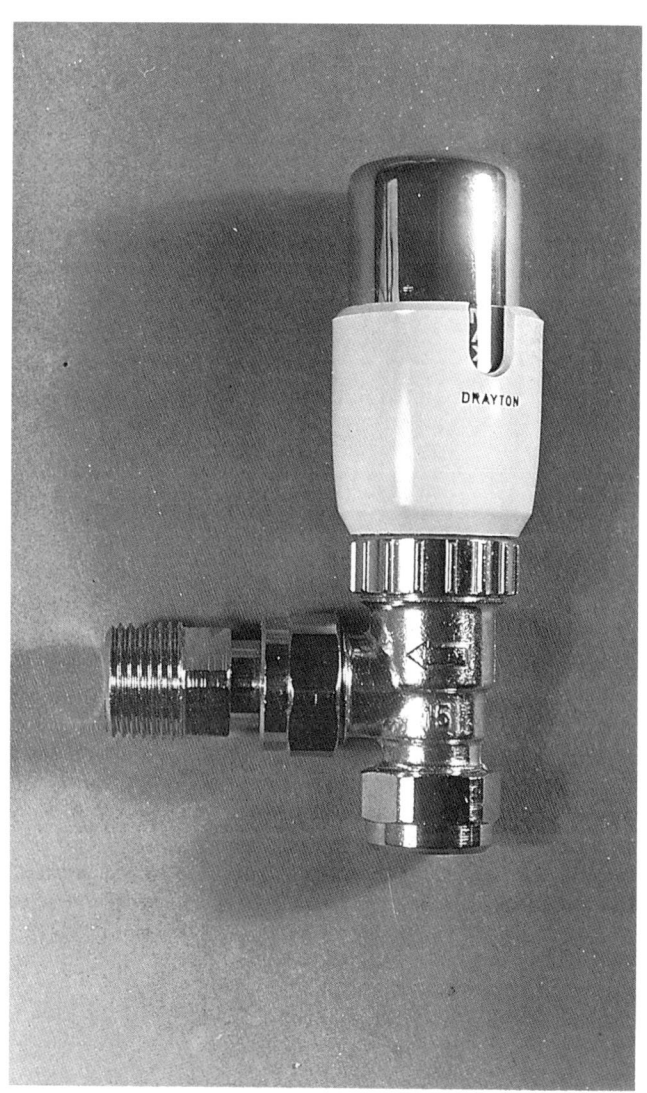

The basic one-way motorised valve, which is electrically operated, is used in conjunction with a thermostat to allow or prevent the flow of hot water through the pipe.

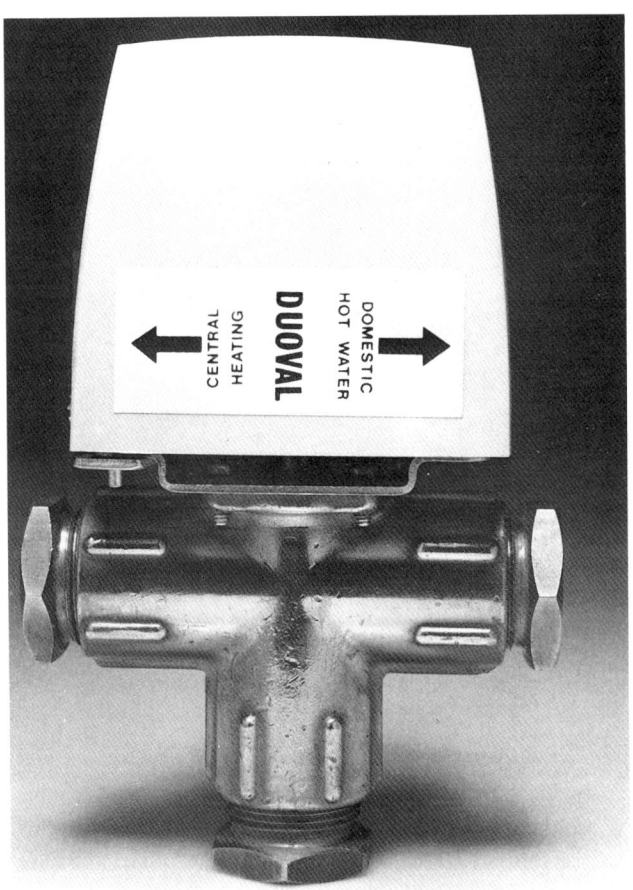

The three-way motorised valve is designed to control two circuits with a selection of options. Each circuit can be controlled so that you can have just one on or both on – or even both off.

particularly important with some wall-mounted gas boilers, where the pump must be able to circulate water through the boiler and prevent damage due to an insufficient amount of water in it.

If a system is installed using two simple valves as described, then bypass pipework must be incorporated to allow the pump to circulate water after the boiler has been shut down (see pages 108–111). When fitting this, you will have a special time-delay circuit in the pump and boiler controls.

Electrical safety

All metal parts of electrically operated items must be earthed to prevent the possibility of the system becoming 'live' as a result of an electrical fault. It is most important that you install a double-pole isolator to protect all the control circuitry, since in many cases after the programmer is turned off some of the circuits to the motorised valves are still live. This can be achieved by wiring all the control circuitry through a 13 amp three-pin plug.

All wiring connected to controls must conform to the IEE regulations. Some thermostats are available that operate on a low voltage for additional safety. If you are considering fitting these, make sure that your central heating system will operate effectively with them.

In a complex electrical circuit, you are advised to get the wiring checked by a competent and qualified electrician to ensure it is correctly and safely installed. It is not unusual to find that where a central heating system is not operating correctly, the fault lies in the wiring of the controls. For example, one reason why a boiler does not start when switched on may be that none of the thermostats in the circuit are calling for heat. Equally two thermostats controlling the same pump may be conflicting – where one is calling for heat and the other is not.

One precaution well worth taking when wiring up a system of controls is to label all the wires and connections. This will make any future maintenance or repair work on the system that much easier to handle.

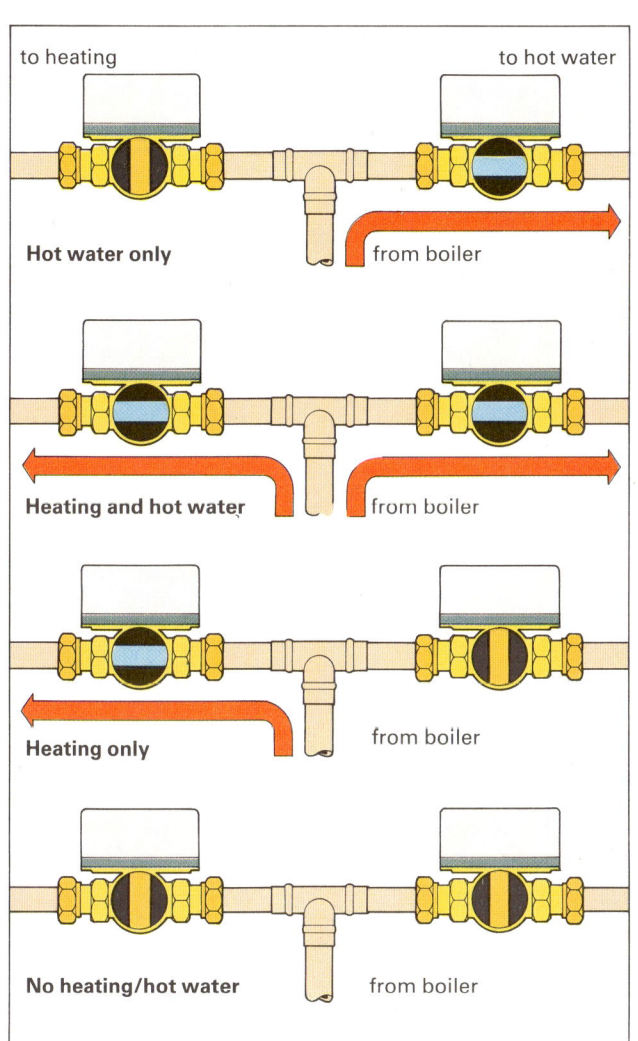

Hot water only

to heating   to hot water

from boiler

Heating and hot water

from boiler

Heating only

from boiler

No heating/hot water

from boiler

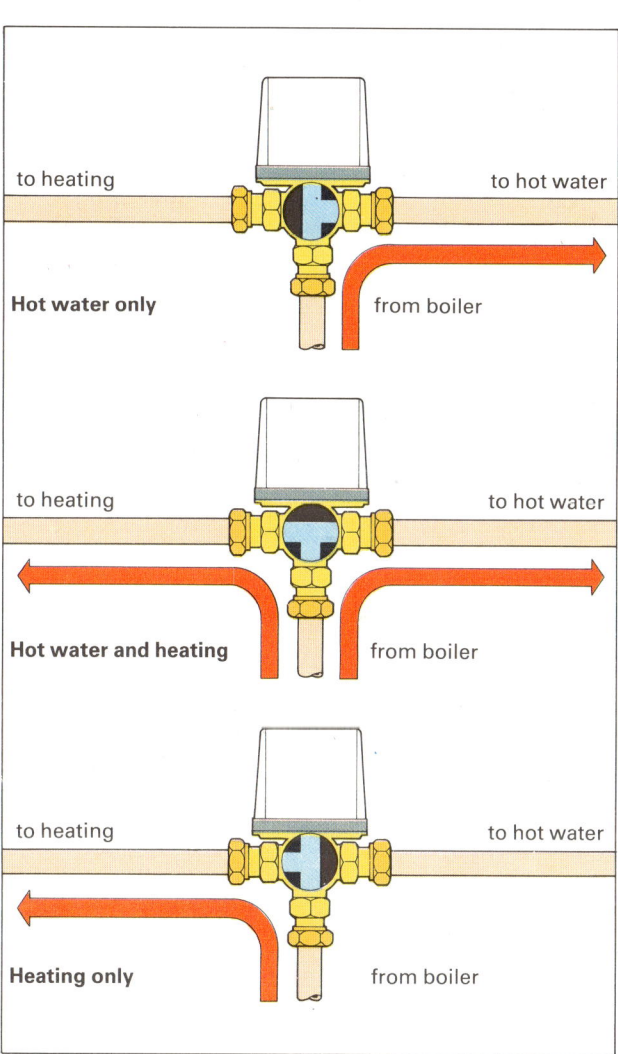

to heating   to hot water

Hot water only   from boiler

to heating   to hot water

Hot water and heating   from boiler

to heating   to hot water

Heating only   from boiler

This set of diagrams illustrates how a domestic hot water and central heating circuit can be controlled by two simple one-way motorised valves, with options ranging from both on to both off. Each valve is linked to its own thermostat.

In this series of diagrams you can see how one more sophisticated valve can control both the domestic hot water and central heating circuits, offering three options. It is impossible to shut off both circuits at the same time.

5 Planning the system

This is obviously the most critical part of the operation, since any mistakes made in the calculations will be very expensive to put right later on. Of course, advice on the type of system and how it should be planned is available through the people from whom you buy your central heating. The information given here is to help you understand what is involved and how the calculations are made. If you are in any doubt, then you should seek expert advice in this all-important preparatory stage of the work.

Calculating heat needed

This is, without doubt, the most important part of the planning of the system, since these calculations will determine the size and output of the boiler, the size and number of radiators needed, as well as the pump size.

Although the calculations are somewhat tedious to work out, they do not require any special mathematical skill – and the method used will be the same, regardless of the type of system you finally decide to install.

As already stated, if you employ a contractor to do the work then he will handle the calculations for you. Equally, some of the larger DIY stores offer a design service and will plan your system based on certain information that you supply them. If you decide to use storage radiators, your local Electricity Board will handle the design and planning.

To calculate your heat requirements, you will need to gather the following information:
- temperature required in each room
- ventilation required in each room
- amount of heat lost or gained through the walls, floor and ceiling of each room
- heat needed for domestic hot water.

When making these calculations, you will need to draw a plan of each floor in the house to be heated, incorporating wall thicknesses and window and door sizes. Measure the height of the ceiling and note the thickness of any loft insulation. You will also need to know the type of floor construction in each room and, where applicable, the gap between the panes of any double-glazed window.

Room temperature The normal recommended room temperatures are as follows:
- lounge/dining room/bed-sit   21°C
- kitchen/bathroom/bedrooms   18°C
- hall/landing   16°C

You may well want to vary these standard temperatures according to your own requirements. For example, an elderly person likely to spend most of the time in the lounge will probably want a slightly higher temperature.

If you are not going to install the boiler in the kitchen, you may wish to increase the temperature level in this room to 21°, depending on how often you use the cooker. Taking into account what additional heating you have in the house, you may want to reduce the temperature level in particular rooms.

Ventilation For health reasons and to prevent rooms from getting stuffy, as well as to reduce the potential level of condensation, the air in each room needs to be changed periodically. This normally happens through the inevitable draughts and gaps through doors, floors, etc and the general movement of people through the house. Unfortunately this air has to be heated as well and so you need to make an allowance for this in your final calculations.

The accepted rates of ventilation of specific rooms are expressed in terms of the number of air changes per hour. These are as follows:
- lounge/bedrooms/bedsits   1
- hall/landing $1\frac{1}{2}$
- kitchen/bathroom/dining room   2

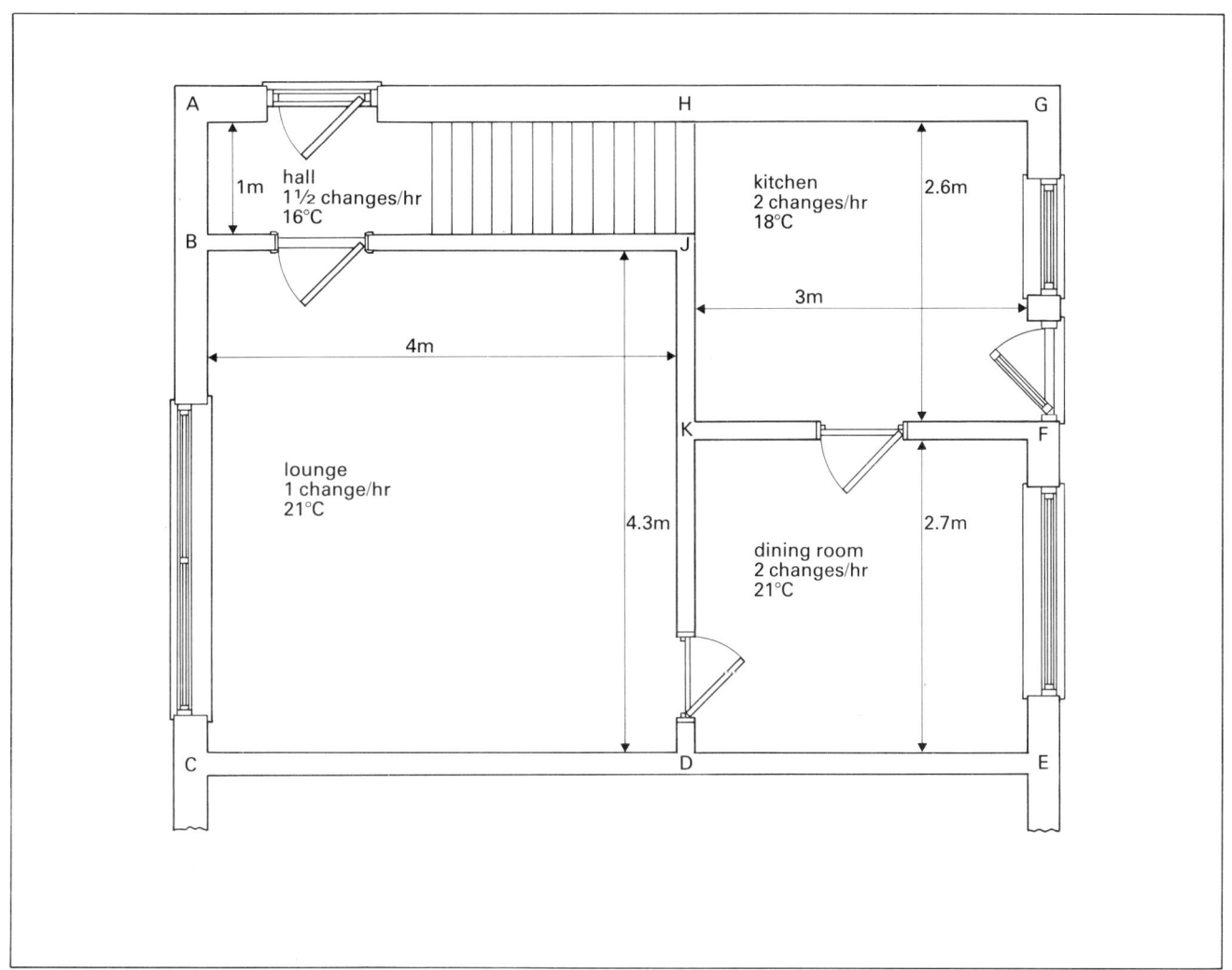

This diagram shows the ground floor layout of a typical semi-detached three bedroom house. For your heat requirement calculations you will only need internal measurements. In addition to the information shown in this typical plan, you will also need to measure the size of the windows and external doors. Another necessary measurement will be the height of the ceiling and you will also require details of the wall and floor materials.

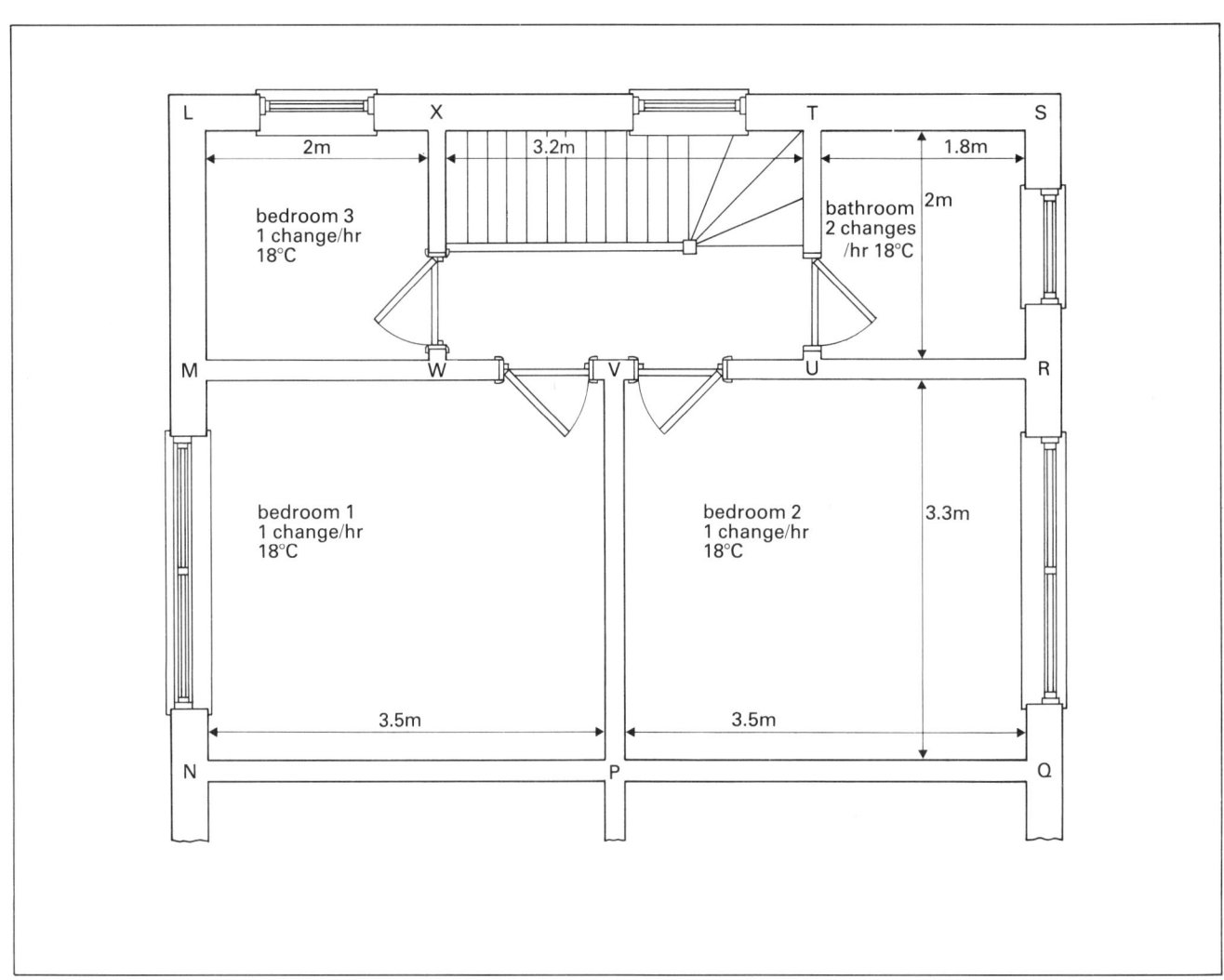

This plan is of the first floor of the same semi-detached house represented in the previous page. Again, for your calculation you will only require internal measurements but you will, as before, need window sizes and also the height of the ceiling in each room. To complete the heat requirement calculations you will have to check on the composition of the intermediate floor and walls, as well as the type of roof and the kind of insulation – and its depth.

These rates should be used as a general guide and, of course, can be varied according to individual requirements and conditions.

To calculate the heat required for ventilation, multiply the volume of the room (in m³) by the number of air changes, the density of air (1.2 kg/m³) and the specific heat capacity of air – or the amount of heat needed to raise 1 m³ of air through 1°C – normally taken as 1010 joules/kg/°C, and the temperature difference. Then divide this by 3600 to get the answer in watts. The formula for this calculation is:

$$\text{heat required (watts)} = \frac{\text{volume of room} \times \text{air changes/hr} \times 1.2 \times 1010 \times \text{temp. diff.}}{3600}$$

Heat losses or gain   Heat will pass through all room surfaces – wall, windows, doors, ceilings and floors – unless the temperature on either side of a particular surface is the same. The heat always passes from the hotter area to the colder area and the rate at which this happens can be measured. A set of standard values for the most common materials is set out opposite.

The rate of heat transmittance is called the U value – expressed in watts/m²/°C. The values shown in the table are for those houses with average exposure to the elements. For the purposes of calculation, the outside temperature is usually measured at $-1°C$. If you live in more exposed conditions, such as Northern England or Scotland, it is as well to work on $-2°C$.

With double-glazed windows, the normal U values quoted do not take into account the heating effect of the sun shining through the window. The Glass and Glazing Federation has now arrived at a set of alternative U values (see Table), which includes this effect and take into account the direction in which the windows face.

To calculate the heat loss through a particular room surface, you will need to multiply the area of the wall (in m²) by the U value of the wall material and the

---

Table of U values

The U values shown here are for properties with average exposure to the elements. If you live in a particularly exposed, windswept situation, you should increase the values for outside surfaces by about 10%.

External walls (plastered internally)

255 mm (9 in) solid brickwork	2.17
brick/50 mm (2 in) cavity/breeze block	1.37
brick/50 mm (2 in) cavity/concrete block	0.96
brick/50 mm (2 in) insulated cavity/breeze block	0.48
brick/50 mm (2 in) insulated cavity/100 mm (4 in) insulating block	0.43
solid/150 mm (6 in) cellular concrete/tile-hung fascia	0.90
timber/framed panels/50 mm (2 in) mineral fibre insulation	0.52

Partition walls

100 mm (4 in) breeze block plastered both sides	2.06
50 mm (2 in) timber framework/12 mm ($\frac{1}{2}$ in) plasterboard each side	1.74

Ground floor

solid/in contact with earth/thermoplastic tiles:
- detached house   0.75
- semi-detached house   0.61
- terraced house   0.48

solid/in contact with earth/underlay and carpet:
- semi-detached house   0.53

- wooden/ventilated/on joists/thermoplastic tiles   0.59
- wooden/ventilated/on joists/underlay and carpet   0.51

Intermediate floor

wooden/on joists/plaster ceiling	1.7
wooden/on joists/plaster ceiling/underlay and carpet	1.2
150 mm (6 in) concrete/50 mm (2 in) screed/underlay and carpet	1.6

External doors

50 mm (2 in) solid wood	2.2
glazed	as for windows

Windows

single glazed	5.6
double-glazed/12 mm ($\frac{1}{2}$ in) gap	3.0
double-glazed/20 mm ($\frac{3}{4}$ in) gap	2.9

With windows, the Glass & Glazing Federation has issued new values which take into account the heat gains that can take place (even in winter) due to the radiation from the sun. This means that the 'effective U value', as it is known, varies depending on the direction in which the window faces. It is equally dependent on the average amount of sunlight to which your house is exposed during the main heating season.

If you use this set of values, the heating requirement for rooms with large windows will be considerably reduced.

Roof

Typical 30° pitch/tiles on battens or boards/sarking felt/plaster ceiling covered with:
50 mm (2 in) blanket insulation	0.55
100 mm (4 in) blanket insulation	0.33
100 mm (4 in) loose-fill insulation	0.47
flat/three layers of felt/chipboard/joists/ plasterboard	1.05
as above with 50 mm (2 in) mineral fibre insulation	0.52
flat/asphalt on screed in 150 mm (6 in) concrete/50 mm (2 in) mineral fibre insulation between screed and concrete	0.85

Effective U values for single glazing

north-facing window	4.36
east/west-facing window	3.62
south-facing window	2.27

Efective U value for double glazing with 12 mm ($\frac{1}{2}$ in gap)

north-facing window	1.92
east/west-facing window	1.28
south-facing window	0.10

If the U values you need are not included here, you should be able to obtain them from the Data Book issued by the Institute of Plumbing, which should be available through your local library.

Table A

Spacings for copper pipe clips		
Nominal size (mm)	Intervals for vertical runs (m)	Intervals for horizontal runs (m)
15	1.9	1.3
22	2.5	1.9
28	2.5	1.9
35	2.8	2.5
42	2.8	2.5

Table B Heat emissions from copper pipes (average conditions)

Nominal size (mm)	Exposed (W/m)	Sleeved or lagged (W/m)
6	17	4
8	23	6
10	29	7
12	34	9
15	43	11
22	63	16
28	77	19
35	92	23

Table C Multiplying factors for conversion of radiator emission values for temperature difference other than 55°C

°C	0	1	2	3	4	5	6	7	8	9
10	0.11	0.12	0.14	0.15	0.17	0.19	0.20	0.22	0.23	0.25
20	0.27	0.29	0.30	0.32	0.34	0.36	0.38	0.40	0.42	0.44
30	0.46	0.48	0.49	0.52	0.54	0.56	0.58	0.60	0.62	0.64
40	0.66	0.68	0.70	0.73	0.74	0.77	0.79	0.82	0.84	0.86
50	0.88	0.91	0.93	0.95	0.98	1.00	1.02	1.05	1.07	1.09
60	1.12	1.14	1.17	1.19	1.22	1.24	1.26	1.29	1.32	1.34
70	1.37	1.39	1.42	1.45	1.47	1.50	1.52	1.55	1.58	1.60

Table D Flow characteristics of water at 75°C
Copper pipes to BS 2871: Part 1: 1971
Pressure loss in N/m² per metre run of pipe

Flow rate (kg/s)	Equivalent heatflow (W/°C)	15 mm	22 mm	28 mm	35 mm
0.010	42	9			
0.016	67	18	3		
0.020	84	27	4		
0.025	105	40	6		
0.030	125	54	8	2.5	
0.035	146	71	11	3.5	
0.040	167	90	14	4	
0.045	188	110	17	5	
0.050	209	132	20	6	
0.055	230	155	24	7	
0.060	251	181	28	8	
0.065	272	209	32	9	
0.070	293	237	36	11	
0.075	314	267	41	12	
0.080	334	299	46	14	
0.085	355	335	51	15	5
0.090	376	370	56	16	6
0.095	397	406	62	18	6.5
0.10	418	445	68	19	7
0.11	460	527	80	23	8
0.12	502	616	93	27	10
0.13	543	709	107	31	11
0.14	585	808	122	35	13
0.15	627	913	137	40	14
0.16	669	1025	154	45	16
0.17	711	1140	171	50	18
0.18	752	1263	190	55	19
0.19	794		210	60	21
0.20	836		230	66	23
0.21	878		250	72	25
0.22	920		271	78	28
0.23	961		294	85	30
0.24	1003		316	91	32
0.25	1045		340	98	35
0.26	1087		362	105	37
0.27	1129		390	113	40
0.28	1170		417	120	42
0.29	1212		443	128	45
0.30	1254		470	135	48
0.31	1296		500	144	51
0.32	1338		528	152	54
0.33	1379			161	57
0.34	1421			169	60
0.35	1463			178	63
0.36	1505			188	66
0.37	1547			197	69
0.38	1588			206	73
0.39	1630			216	76
0.40	1672			226	80
0.41	1714			237	83
0.42	1756			247	87
0.43	1797			257	90
0.44	1839			268	94
0.45	1881			280	98
0.46	1923			291	102
0.47	1965			302	106
0.48	2006			313	110
0.50	2090			338	119
0.52	2174			362	127
0.54	2257			387	136
0.56	2341				145
0.58	2424				154
0.60	2508				163
0.62	2592				174
0.64	2675				183
0.66	2759				194
0.68	2842				204
0.70	2926				215
0.72	3010				227
0.74	3093				239
0.76	3177				250
0.78	3260				262
0.80	3344				274 *1*
0.82	3428				286 *m/s*

Table E Maximum load-carrying capacities of copper pipes

Nominal size (mm)	Mass of water (kg/s)	W 1½°C drop	W 15°C drop	Velocity (m/s)
6	0.023	960	1 440	1.5
8	0.049	2 050	3 075	1.5
10	0.085	3 550	5 325	1.5
12	0.126	5 270	7 905	1.5
15	0.17	7 110	10 665	1.2
22	0.31	12 960	19 440	1.0
28	0.52	21 740	32 610	1.0
35	0.80	33 440	50 160	1.0

Table F Flow characteristics of hot water in microbore pipes

Pressure loss in N/m^2 per metre run of pipe

Flow rate (kg/s)	Equivalent heat flow (W/°C)	6 mm	8 mm	10 mm	12 mm
0.007	29	810	182	46	21
0.008	33	975	210	55	28
0.009	38	1165	245	68	34
0.010	42	1395	285	81	40
0.011	46	1625	325	94	47
0.012	50	1885	360	107	53
0.013	54	2165	405	120	60
0.014	59	2475	460	133	66
0.015	63	2800	525	145	73
0.016	67	3155	595	164	79
0.017	71	3535	665	180	88
0.018	75	3930	740	195	97
0.019	79	4350	815	215	107
0.020	84	4800	903	234	113
0.021	88	5330	990	253	120
0.022	92	5780	1075	273	126
0.023	96	6210	1165	293	133
0.024	100		1260	313	139
0.025	105		1350	338	146

Flow rate (kg/s)	Equivalent heat flow (W/°C)	6 mm	8 mm	10 mm	12 mm
0.031	130		2010	520	223
0.037	155		2755	754	313
0.043	180		3585	1030	413
0.049	205		4485	1325	536
0.055	230			1670	665
0.061	255			2030	795
0.067	280			2410	925
0.073	305			2830	1080
0.079	330			3280	1235
0.085	355			3780	1408
0.090	376				1547
0.095	397				1710
0.100	418				1860 *1.2 m/s*
0.105	439				2030 *m/s*
0.110	460				2220
0.115	481				2365
0.120	502				2540
0.126	527				2785 *1.5 m/s*

difference in temperature between either side of the wall. Doors and windows are treated separately. The formula for this calculation is:

heat loss (watts) = area (m²) × U value × temp. diff.

The example below clarifies these calculations. The diagram shows the ground and first floor of a typical semi-detached house. For the purposes of the example, the following information is given:

● adjacent semi-detached house already heated to similar desired temperatures in the adjoining rooms
● windows double-glazed with a 12 mm ($\frac{1}{2}$ in) air gap
● outside walls brick cavity with foam insulation
● loft has 100 mm (4 in) glass fibre insulation
● the ceilings are all 2.8 m high
● both sets of floors suspended, wooden and carpeted
● all doors 2 m high and 750 mm wide (internal doors may be ignored in the calculations).

From these calculations you can take the total heat requirement for each room, as follows:

lounge   − 357 + 742   = 1099 watts
kitchen   − 279 + 282   = 561 watts
bathroom   − 129 + 200.2 = 329.2 watts
bedroom 1   − 207 + 253   = 460 watts

Add together the values of all the rooms, calculating the other rooms in the same way, to give the total central heating requirement for the house.

Heat for domestic water   In order to calculate this, you will need to know approximately how much hot water the household is likely to need. As a guide, a bath uses about 68 litres (15 gallons) of hot water, mixed with a similar amount of cold. This means that a hot water cylinder with a capacity of 136 litres (30 gallons) should supply enough hot water for two baths – the requirement of the average family. If you have a large family or household – or you need to supply two bathrooms – consider a larger hot water cylinder.

Normally the minimum time required to heat 136 litres (30 gallons) of water is two hours. Bear in mind that the quicker you want to heat the water, the more it will cost. Domestic water is normally heated to a temperature of 60°C – and must not, for safety reasons, be heated to more than 65°C.

To work out the heat requirement, multiply the volume of water (in litres) by the specific heat capacity of water – that is the amount of heat needed to raise 1 kg of water through 1°C, taken as 4.2 kilojoules/kg/°C – and the difference in temperature between hot and cold water. To get the answer in kilowatts, divide this by 3600. This will give you the amount of heat required to heat the water from cold in one hour. The formula is:

$$\frac{\text{heat required}}{\text{(in kW)}} = \frac{\text{volume of water (1)} \times 4.2 \times \text{temp. diff.}}{3600}$$

The heat required for ventilation is calculated as follows:

lounge $\dfrac{(2.8 \times 4.0 \times 4.3) \times 1 \times 1.2 \times 1010 \times (21 - (-1))}{3600} = 357$ watts

kitchen $\dfrac{(2.8 \times 3.0 \times 2.6) \times 2 \times 1.2 \times 1010 \times (18 - (-1))}{3600} = 279$ watts

bathroom $\dfrac{(2.8 \times 2.0 \times 1.8) \times 2 \times 1.2 \times 1010 \times (18 - (-1))}{3600} = 129$ watts

bedroom 1 $\dfrac{(2.8 \times 3.3 \times 3.5) \times 1 \times 1.2 \times 1010 \times (18 - (-1))}{3600} = 207$ watts

The heat losses and gains through walls, floors and ceilings are calculated as follows			loss	gain
				watts
lounge	wall BC – $((2.8 \times 4.3) - (2.7 \times 1.45)) \times 0.48 \times 22$	=	85.8	—
	wall CD –	=	—	—
	wall DK –	=	—	—
	wall KJ – $(2.8 \times 1.6) \times 2.06 \times 3$	=	27.6	—
	wall JB – $(2.8 \times 4) \times 2.06 \times 5$	=	115.4	—
	window – $(2.7 \times 1.45) \times 3 \times 22$	=	258.4	—
	ceiling – $(4.0 \times 4.3) \times 1.2 \times 3$	=	61.9	—
	floor – $(4.0 \times 4.3) \times 0.51 \times 22$	=	192.9	—
	total losses	=	742.0	
kitchen	wall GF – $((2.8 \times 2.6) - ((1.0 \times 1.0) + (2.0 \times 0.75))) \times 0.48 \times 19$	=	43.6	—
	wall GH – $(2.8 \times 3.0) \times 0.48 \times 19$	=	76.6	—
	wall HJ – $(2.8 \times 1.0) \times 2.06 \times 2$	=	11.5	—
	wall JK – $(2.8 \times 1.6) \times 2.06 \times 3$	=	—	27.7
	wall FK – $(2.8 \times 3.0) \times 2.06 \times 3$	=	—	51.9
	window – $(1.0 \times 1.0) \times 3 \times 19$	=	57.0	—
	door – $(2.0 \times 0.75) \times 3 \times 19$	=	85.5	—
	ceiling –	=	—	—
	floor – $(3.0 \times 2.6) \times 0.59 \times 19$	=	87.4	—
	total losses	=	282.0	
bathroom	wall SR – $((2.8 \times 2.0) - (1.0 \times 1.2)) \times 0.48 \times 19$	=	40.1	—
	wall ST – $(2.8 \times 1.8) \times 0.48 \times 19$	=	46.0	—
	wall TU – $(2.8 \times 2.0) \times 2.06 \times 2$	=	23.1	—
	wall RU –	=	—	—
	window – $(1.0 \times 1.2) \times 3.0 \times 19$	=	68.4	—
	floor –	=	—	—
	ceiling – $(2.0 \times 1.8) \times 0.33 \times 19$	=	22.6	—
	total losses	=	200.2	
bedroom 1	wall MN – $(2.8 \times 3.3) - (2.4 \times 1.2) \times 0.48 \times 19$	=	58.0	—
	wall NP –	=	—	—
	wall PV –	=	—	—
	wall MV –	=	—	—
	window – $(2.4 \times 1.2) \times 3.0 \times 19$	=	164.2	—
	floor – $(3.5 \times 3.3) \times 1.2 \times 3$	=	—	41.6
	ceiling – $(3.5 \times 3.3) \times 0.33 \times 19$	=	72.4	—
	total losses	=	253.0	

Ideally the pipes between the boiler and the hot water cylinder should be lagged to prevent any heat loss. If not, you must make an allowance to compensate for such loss. From Table B you can obtain the heat loss per metre of pipe run. Multiply this loss by the length of pipe in metres and add the result to the required heat amount.

Your hot water cylinder should also be lagged, although even with this insulation heat is still lost at a rate of about 145 watts per hour. So you must also add this to the heat required amount.

You will now have the figure for the total heat required for one hour. If you are estimating this over a two-hour period, simply divide by two.

In the example shown above, the domestic hot water requirement is as follows:

suitable cylinder size	136 litres
desired water temperature	60°C
time to heat from cold	2 hours
initial water temperature	10°C

Therefore:

$$\text{heat for 1 hour} = \frac{140 \times 4.2 \times (60 - 10)}{3600} = 8.17\,\text{kW}.$$

The pipes between the boiler and the hot water cylinder are to be lagged, so potential heat loss here can be ignored. The loss of heat through the cylinder jacket is 145 watts/hr, so the heat required for a one-hour period is:

$$8.17 + 0.145 = 8.315\,\text{kW}$$

For a two-hour heating period, therefore, the heat required is 4.157 kW.

Having reached this stage of the calculations, you now need to look at the radiators.

Types of radiator
Basically there are four main types of radiator – panel, skirting and convector radiators and bathroom towel rails.

Panel radiator   This is the most commonly used type of radiator and different heights and lengths are available to suit the location and output required. Make sure this radiator is made to BS 3528: 1977.

The basic type has a single panel. To increase the heat output per square metre, a double panel radiator is available. There is a further variation in design, which incorporates the addition of special convector fins on the back that can increase the output by up to 50 per cent. As the design implies, this type is called a convector radiator, not to be confused with convector heaters (see below).

The panel radiator is hung on brackets screwed to the wall. There are four tappings for connection to the central heating system. The normal method of connection involves using the lower tappings for the flow and return pipes; one of the top tappings is blanked off and the other fitted with an air-release valve.

Panel radiators can only be used where the temperature of the water in the central heating system does not exceed 80°C. They are totally silent in operation and need no maintenance other than the occasional redecoration.

Skirting radiator   As the name implies, this type of radiator is fitted in place of the skirting board along the bottom of the wall. It consists of a heat exchanger encased in pressed steel. The heat exchanger is usually a copper pipe fitted with vertical square or rectangular fins.

Air enters through the bottom of the casing and rises by convection, picking up the heat as it passes through the fins. The hot air then flows through a damper and into the room. The damper can be used to

Panel radiators, such as the single ones shown here, come in a range of shapes and sizes. Double panel models for greater heat output are also available. Depending on the wall space available to you when you fit a radiator, you should in preference fit it along, rather than up, the wall to ensure the greatest possible spread of heat. When you calculate the size of radiator you need, always estimate for one slightly larger to allow for spare capacity.

As you can see from this room, skirting radiators blend well into the general shape and decoration when fitted in place at floor level. Apart from looking unobtrusive, they are particularly useful where space is at a premium.

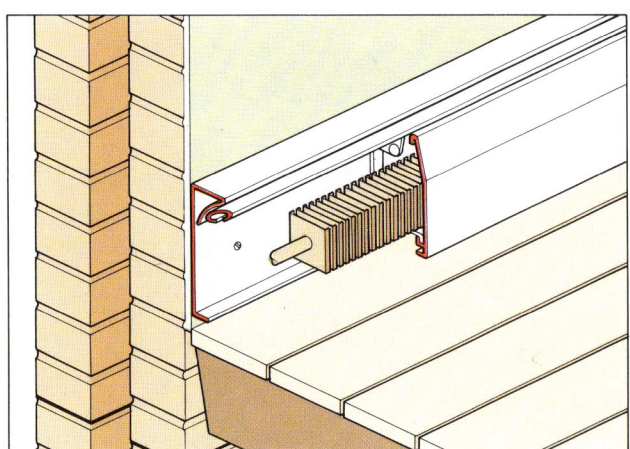

This diagram illustrates how a typical skirting radiator is constructed. Inside the casing is the heat exchanger, which runs the length of the heater. It is made up of a series of fins wrapped round a copper tube through which the hot water flows.

regulate the flow of air and therefore the heating output of the radiator. The surface area of the fins can be corrugated to increase efficiency.

The flow pipe is connected to the tapping at one end and the return pipe to the tapping at the other end. Some types have support brackets (see diagram) fitted so that the return pipe can be run back inside the radiator above the heat exchanger. This enables both feed and return pipes to come from the same end of the radiator if required.

One of the advantages of this type of radiator is that the casing and the heat exchanger come separately. This means you can fit the casing to the desired length along a wall to improve the overall look, but use just the required length of heat exchanger and position it anywhere inside the casing.

Provided that the feed and return pipes are out of reach and not accessible to the touch, the water temperature can be more than 80°C, if required. This type of radiator will also perform very efficiently at 80°C, enabling it to be used in conjunction with panel radiators if required.

The skirting radiator is completely silent in operation, but it does require some general maintenance. Periodically you will need to remove the casing and clear away dust and fluff that will have collected between the fins in the heat exchanger. You can do this with a vacuum cleaner and a damp cloth.

Convector radiator   This type incorporates an electric fan, which blows air past a heat exchanger into the room. The designs of heat exchanger do vary, but are usually similar to those used in skirting radiators. The basic difference is that the pipes run back and forth through the exchanger fins.

The fan has several settings so that you can regulate the air flow and therefore the heat output. When a fan is not operating, heat is given off – at a reduced rate – through convection. Normally a thermostat is fitted which switches the fan on and off as required. This is usually built into the heater, although you can get a remote thermostat (see page 69).

The main advantage of the convector radiator is that it takes up a lot less space than a conventional radiator and is therefore particularly useful in those rooms where wall space is at a premium – such as in a kitchen.

The convector radiator can be used with water temperatures at 80°C or more, provided that the connecting pipes are out of reach. It is not completely silent in operation, the noise level being similar to that of a fan heater. Also, it requires a properly installed supply of electricity (about 100 watts) and the fan and heat exchanger will need to be cleaned periodically.

Bathroom towel rail   This is a common bathroom fitting which should preferably be connected to the domestic hot water system. The reason for this is that damp towels need drying all the year round – and not just when the central heating is switched on.

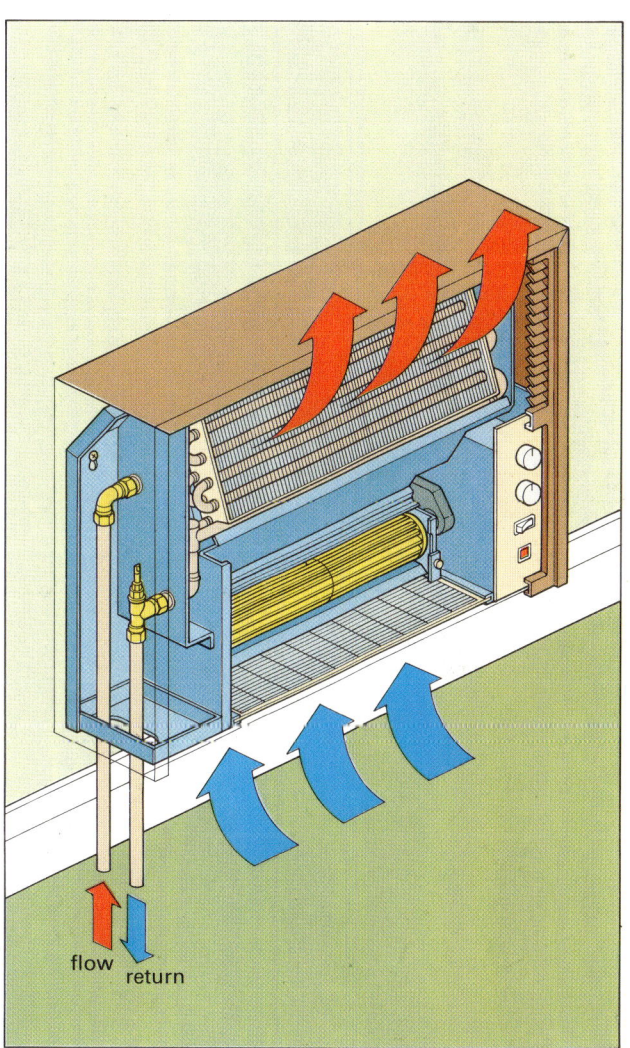

flow return

The fan convector heater, details of which are shown here, has a high efficiency heat exchanger over which air is drawn at a variable rate, depending on the speed of the electric fan, and then let out at the top to heat the room.

Towel rails are useful and valuable additions to the bathroom. They should be connected, preferably to the domestic hot water system so that towels can be dried throughout the year and not simply during the colder months when the central heating is switched on.

A kick-space fan convector is a particularly convenient solution to the problems of heating the kitchen area. The type of heater is designed to fit neatly into the kick-space *area between the bottom of a cupboard door and the floor into the plinth. It needs to be plumbed in to a pump-assisted wet central heating system.*

Particularly with central heating fired by solid fuel, this provides a useful means of using up excess heat from the boiler.

Different models and styles are available, depending on the size of the bathroom and the wall space available. The range varies from wall-mounted rails to the larger floor-mounted ones that incorporate a radiator and come with either a chrome or gold-plated finish to suit individual bathrooms.

Siting radiators

Although there can be no definite rules as to where to place radiators, there are several points worth bearing in mind which will influence their positioning.

With a single-glazed window, where you will get a false draught caused by the cooling effect of the pane, you can to an extent counteract this by positioning a radiator below the window. With a double-glazed

Fan convector heaters are now available in slimline designs that fit neatly into most areas in the home. Normally the water that passes through them is hotter than for normal radiators and the pipework should therefore be hidden.

window this is not so important. Where you do install a radiator under a window, make sure when the curtains are drawn that they hang behind it. If not, much of the heat emitted from it will pass up behind the curtains and out through the window.

A certain amount of heat from a radiator will be lost into the wall behind. Where panel radiators are concerned, this can be counteracted by fitting a reflective aluminium foil panel behind each radiator to throw the heat back into the room. This is particularly important with an external wall. With an internal wall, any heat lost will, of course, pass through to the adjoining room.

Another obvious, yet often neglected, fact is that

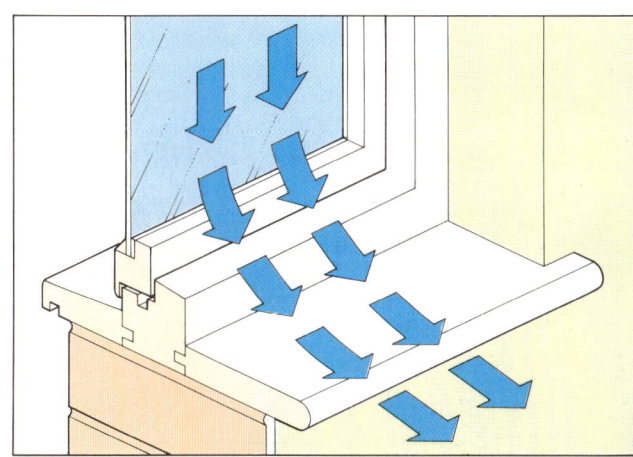

From this diagram you can see how a false draught can be given off from a single pane of glass when it is cold. As the warm air from the room hits the window and cools, it falls down and then comes back into the heart of the room.

large pieces of furniture can interrupt the flow of heat into a room. Therefore you should avoid placing sizeable items of furniture directly in front of a radiator.

When fitting a radiator to a plain wall. the convection of air behind and above the radiator will over a period of time lead to a dirty stain on the wall. One way of preventing this from happening is to fit a shelf above the radiator. This has the effect of deflecting the heat off the wall and into the room. This will cause a slight loss in the efficiency of the radiator, but you can make allowances for this when calculating the optimum size of radiator (see below).

One very important consideration when selecting a site for each radiator is the length and route of pipe run required to supply that radiator. As a general rule, you should keep pipe runs as short as possible. This is important for several reasons – appearance, efficiency and cost.

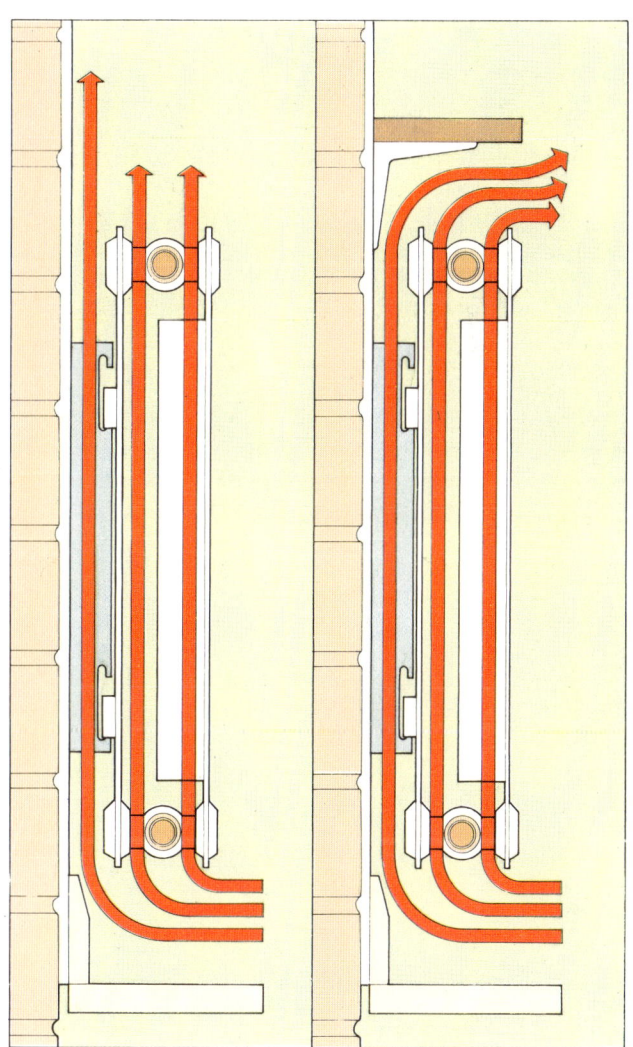

There are distinct advantages of fitting a shelf above a radiator. Not only will it deflect the hot air away from the wall and into the room, but it will also prevent deposits of dirt spoiling the wall coverings or decoration.

Where you have to site radiators in adjoining rooms, as far as possible these should be back to back to simplify the plumbing. When considering the feed and return pipes, these should never be embedded in a concrete floor or plastered into the wall, since it will make further maintenance impossible without a lot of trouble and will also result in problems due to the expansion in the pipework.

Depending on the wall space available, it is much better practice to install a long, low radiator rather than a short, tall one. You will find that by spreading the area of heat emission along rather than up a wall, you will distribute the heat more evenly.

In most average size rooms you will probably only need one radiator. In larger rooms – and where they are irregularly shaped (for example, an L-shaped room) – two or more will distribute the heat efficiently.

Although there are several considerations that will determine the exact location of the radiators, the most common position for them is under the window.

Fitting a kick-space fan convector

One particularly convenient solution to the problem of heating the kitchen area of a home, especially when it is of a compact, modern design with limited floor and wall space, is to install a kick-space fan convector.

This type of heater, which fits neatly in the kick-space area between the bottom of a cupboard door and the floor into the plinth, needs to be plumbed in to a pump-assisted wet central heating system.

The heater needs to be sited so that the flow of warm air from it is into the centre of the room. Do not aim it towards a door. The other determining factor will be where you can most conveniently route the flow and return pipes to be connected to the heater.

Planning the work   You need a fair amount of preparatory work before tackling the job of installing the heater. This will include clearing everything out of the cupboard under which you want to fit the unit and ensuring you have the best possible access for connecting up and running the necessary pipework and making the electrical connections. You will find the work a lot easier if you remove the cupboard doors as well.

Before connecting up the existing central heating system, you must switch off the boiler and then drain the primary circuit. To do this, find the draincock at the lowest point in the circuit and fit to this some hosepipe long enough to take the water away safely. Before you turn on the draincock, turn off the main supply to the feed and expansion tank in the loft, otherwise the sytem will keep filling itself up.

Make sure that you turn off the main electricity supply before you connect up any of the wiring.

Installing the heater   Remove the floor in the cupboard above the required position of the heater, if you can. Otherwise you will have to cut out a section large enough to fit the heater comfortably. In this case, drill a few holes and then insert a padsaw to start the cut, until you can fit the blade of a panel saw. Or use a jig saw.

Having removed the cupboard floor or section of it, make sure that it can be lifted again easily after the installation should any maintenance on the heater be necessary later on. You should be able to remove the kickboard at the front of the cupboard base without too much trouble and here you will need to cut out a suitably shaped hole to fit the grille of the heater.

Having prepared the cupboard floor and plinth, you can fix the heater in place by screwing it down to the kitchen floor under the cupboard. You may need to use wooden blocks on which to mount the heater.

Connecting the heater   You now need to make the necessary connections from the heater to the existing central heating system. Where you tap in the flow and return pipes will depend on the location of the existing pipework. Most circuits with radiators incorporate a two-pipe system, with hot water from the boiler flowing to all the radiators via one pipe, while the radiators return the cooler water via another pipe.

To tap into each of these pipes, you can use one of several T-type fittings currently on the market that are designed for simple DIY installation and will include their own fitting instructions. You will find it easier to balance the heating system if you also fit a lockshield valve into the circuit.

Keep the pipe runs to and from the heater as straightforward as possible and avoid sharp bends and elbow fittings. Because of the confined space in which you will be working, you will find it a lot easier to use compression-type fittings. In any case, these must be used on the final connections to the heater. If your heater is at the lowest point on the heating circuit, fit a draincock on the bottom section of pipework.

Wiring up the heater   The electrical connections to the remote controller and heater should be made from a 2 amp fused spur off the existing power circuit or, easier still, via a fused plug in a socket outlet. Fit the controller to a convenient part of the wall, with the cable running straight down from it in a channel chased into the plasterwork. Other sections of cable can be fed through relevant wall units to the heater.

If you are in any doubt about the electrical connections, then you must consult a qualified electrician for advice.

Checking the system   Once you have completed the installation and checked that all your connections are sound, slowly refill the primary circuit and pay special attention to all the new joints in the pipework for possible leaks. Remember to top up this circuit with corrosion inhibitor as you refill it.

Switch on the boiler and check the temperature of all the radiators on the same circuit as the new heater.

You may find that they remain cold. Try bleeding them to release any air trapped inside.

If this fails to work, the chances are that the hot water in the system is taking an easier route through the new heater rather than the radiators. To correct this, screw down the lockshield valve until the radiators reach their normal temperature and the return pipe from your new heater is still warm. If you have not fitted a separate lockshield valve, you will have to adjust all the others in the circuit.

Take a final check on the electrical circuit from the heater to ensure that the controller is working correctly. Remember that because of the low-limit thermostat setting, the fan in the heater will not operate if the water temperature is too low.

Finally lag the new pipe runs and then replace the kickboard and cupboard floor and rehang the doors.

Calculating radiator size

Taking each area of the home in turn, the first piece of information you will need is the heat requirement for the room. You will also want details of the heat emission from the range of radiators available, which you can get from a supplier's brochure. This is given in watts per radiator or watts/m^2 for a range of radiators. The figures given assume a difference of 55°C between the average temperature of the water in the radiator and the required room temperature.

The normal drop in temperature estimated for water passing through a radiator in a pumped system is 10°C. Thus in a system where the boiler thermostat is set at 80°C, the water leaving a radiator will be at a temperature of 70°C, giving an average temperature from the radiator of 75°C.

When you are fitting a radiator in a lounge, where the temperature required is 21°C, the temperature difference will be 54°C (75°C − 21°C). Likewise in a hall it will be 59°C (75°C − 16°C). In the lounge the standard emission estimate must be reduced slightly since the temperature drop is less than 55°C, while in the hall this figure should be increased since the drop in temperature is more than 55°C.

You can do this by multiplying the standard emission figure by a factor taken from Table C. For example, a standard radiator has a heat emission of 905 watts. In the circumstances quoted above, the emission calculation would be:

lounge $905 \times 0.98 = 887$ watts
hall $905 \times 1.09 = 986$ watts

Panel radiators will not be affected significantly if covered with normal household paints, whereas you will reduce the heat emission if you use metallic paints. If you fit a shelf above the radiator, you should increase the required heat emission of that radiator by about 10 per cent to compensate for lack of efficiency.

Any pipes not visible in the room should be lagged, so you can ignore any heat loss from these. Where you are unable to hide the pipework and therefore cannot lag, you can add the heat emitted by these pipes to that from the radiator. Table D gives the normal heat emission from copper pipes in watts/metre.

With skirting radiators, the system is different. Manufacturers supply details of the emission obtained per metre for various water and room temperatures. They also give the drop in temperature that takes place at different water flow rates. If you combine panel and skirting radiators in the same circuit, you can assume the average water temperature will be 75°C.

The specifications for fan convector heaters include the emission figures at various temperatures, as with skirting radiators. These values are, however, dependent on the position of the selector switch. You are advised to use only the low or medium figures in your design calculations, since this will leave the higher outputs available when you need to warm up a room quickly or when there are cold conditions.

Also given in the specifications are the flow resistances through the heater and you will need these when making your calculations for the pipe and pump sizes (see below). Towel rail heaters are specified in a similar way to panel radiators.

One room that needs special consideration is the kitchen, particularly when the boiler is fitted there, since some heat will be emitted from that. Manufacturers quote figures for heat emission and you will need to deduct these from the heat requirement in the kitchen before calculating the radiator size. In many cases, you may may find it is not necessary to have a radiator at all in the kitchen.

It is most unlikely that your heat requirement will coincide exactly with specific radiators and their output. So always select a radiator with heat emission slightly higher than that required. This way there will always be the capacity for sufficient heat – and if the radiator is putting out too much, you can always turn it down or readjust the thermostat.

Designing the pipe circuit

There are several different methods you can use to run pipes to and from radiators. The simplest is the one-pipe system (see diagram). However this suffers two disadvantages. Firstly the pipe size may have to be large to accommodate the flow of water. Also water to the radiators on the end of the circuit will be supplied at a lower temperature and so these radiators will be less efficient than the rest.

For these reasons the one-pipe system is not normally used as the main system, It can, however, be used with skirting radiators where just one floor is to be heated. These radiators are connected in a ring (see diagram) and the calculations on the length of heating should take into account the progressive reduction in heat emission due to the falling temperature of the water as it passes through the system.

The two-pipe system (see diagram) is more commonly used. Here the flow from the boiler to the radiators is through one pipe – or set of pipes – and the

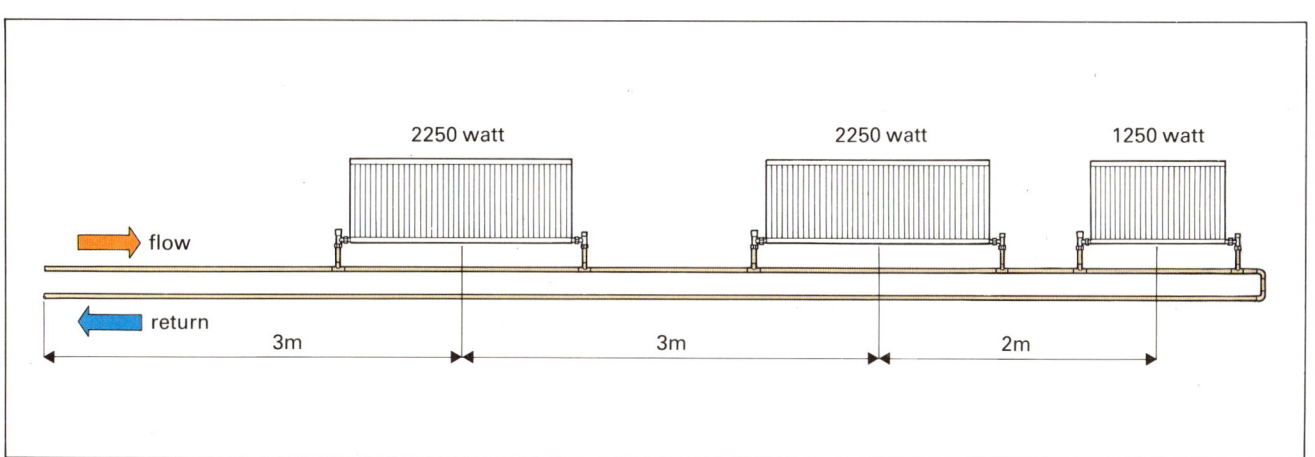

Here you can see how the one-pipe system operates in a central heating circuit feeding a number of radiators. As the hot water circulates, it is fed through each radiator in turn along a single pipe until it has flowed through all the radiators on the circuit and then returns to the boiler to be reheated. Unfortunately with this system the last radiators on the pipe run are supplied with cooler water than the earlier ones and are therefore less efficient.

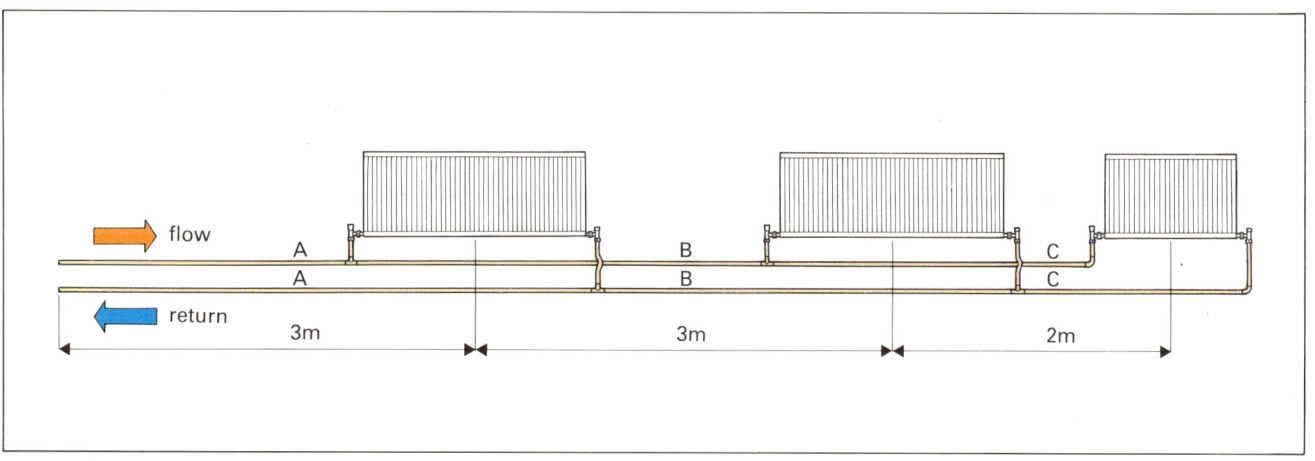

With the two-pipe system, each radiator has a flow and return pipe fitted. This means that the cooler water is returned directly to the boiler and not via the other radiators in the circuit. The benefit of this type of system, which is the one most commonly used in central heating installations, is that all the radiators are able to operate more efficiently. Although twice the pipework is required, this can be smaller and therefore easier to install.

cooler water returns from the radiators through another pipe. Although with this system you will have to use more pipes, usually these can be smaller and therefore easier to install.

You will have to control the flow of water through each radiator separately, otherwise the water will take the route of least resistance and bypass the other circuits. To achieve this, you have to fit special controls, called lockshield valves, to each radiator. These must be adjusted when the system is finally installed and balanced (see pages 111–113) to ensure an even flow of water.

You can run pipes along skirting boards, underneath wooden floors, above ceilings or down the corners of rooms. You should not, as already mentioned, bed them in concrete floors or plaster them into walls. Where pipe runs are not through a heated area – such as a room – you must lag them to prevent heat loss and the possibility of freezing.

Calculating pipe sizes

Obviously there is a limit to the amount of water that can pass through a pipe of a given size. Noise (due to high flow rates) and pump pressure are two limiting factors. To minimise calculations, the amount of water is expressed as the equivalent heat flow rate. This is the amount of water required to supply heat to a radiator divided by the drop in temperature inside the radiator (that is, 10°C). Thus, for a radiator with a heat requirement of 4000 watts, the equivalent heat flow is:

$$\frac{4000}{10} = 400 \text{ watts}°\text{C}.$$

Table E shows the pressure loss in the pipe (in Newtons/m^2) for the equivalent heat flow in different size pipes. The table stops where the speed of the water would exceed the accepted maximum. For example, a heat flow of 750 watts/°C will require 22 mm diameter pipe, not 15 mm.

To calculate the pipe size required for a one-pipe system, all of the water has to pass through all of the pipes. You will first need to work out the total heat emission. This is achieved by adding together the heat emissions from all the radiators and any losses from the pipes. Since you do not yet know which pipe size you will be using, you should assume a heat loss of one-third the total emission. Thus the total heat emission is:

$$(2250 + 2250 + 1250) + \frac{(2250 + 2250 + 1250)}{3}$$
$$= 7667 \text{ watts.}$$

Table E gives the maximum recommended heat flow through various size pipes. From this you can see that 15 mm diameter pipe will be sufficient.

Table B (see page 61) gives the heat losses for exposed pipes. Lagged pipes, as already stated, do emit some heat, but for the purposes of this calculation the amount can be ignored. You can now obtain a more accurate figure for the total heat emission, as follows:

radiator loss   = 2250 + 2250 + 1250 = 5750 watts
pipe losses   = 16 m of 15 mm pipe at 43 watts/m*
    = 688 watts
total emission = 6438 watts
(* taken from Table D)

Therefore the equivalent heat flow is:
$$\frac{6438}{10} = 644 \text{ watts/}°\text{C}$$

From Table D (see page 62), you can see that with 644 watts/°C flowing through a 15 mm pipe, there is a pressure loss of approximately 990 Newtons/m^2. To work out the total pressure loss, you have to multiply this figure by the total length of pipe and add one-third. The one-third is to allow for additional losses in pressure from such things as fittings and elbows in the pipework. The calculation is as follows:

total effective pipe length $= 16 + \frac{16}{3} = 21.3 \text{ m}$

total pressure drop $= 21.3 \times 990$
$= 21087 \text{ Newtons/m}^2.$

In a two-pipe system using similar radiators (see diagram), the calculations are as follows:

For pipes A, which have to carry the total emission of all three radiators:

total equivalent heat flow $= \frac{6438}{10} = 644$ watts/°C

With 15 mm pipe, the pressure loss is 990 Newtons/m² and the effective pipe length is 8 m $(6 + \frac{6}{3})$. Therefore the pressure loss in pipes A is:

$8 \times 990 = 7920$ Newtons/m².

For pipes B, the heat emission from the radiators is 3500 watts (2250 + 1250) and the heat loss from pipes is 430 watts (10 × 43). Therefore the total heat emission is 3930 watts. The equivalent heat flow is:

$\frac{3930}{10} = 393$ watts/°C.

With a heat flow of 393 watts/°C through a 15 mm pipe, the pressure loss is 400 Newtons/m². With the effective pipe length at 8 m, the pressure loss in pipes B is:

$8 \times 400 = 3200$ Newtons/m².

For pipes C, the heat emission from the radiators is 1250 watts and the heat loss from the pipes is 172 watts (4 × 43). Therefore the total heat emission is 1422 watts. The equivalent heat flow is:

$\frac{1422}{10} = 142$ watts/°C.

With a heat flow of 142 watts/°C through a 15 mm pipe, the pressure loss is 70 Newtons/m². With the effective pipe length at 5.3 m $(4 + \frac{4}{3})$, the pressure loss in pipes C is:

$5.3 \times 70 = 371$ Newtons/m².

Therefore the total pressure loss for the circuit is as follows:

pipes A + pipes B + pipes C = 7920 + 3200 + 371
$$= 11491 \text{ Newtons/m}^2.$$

From this you will notice that there is a much lower pressure loss than with a single-pipe system.

Taking this example, you can easily see that if radiator 3 needed to be larger, then pipes A would have to be larger, since the equivalent heat flow would then exceed the maximum flow rate for 15 mm pipes. However, this would not affect pipes B and C, which would still be adequate for the rest of the circuit. In fact, in this case, pipes B and C could be reduced in size, although normally in a small bore system 15 mm pipe is the smallest used.

From the information given above, you should be able to design a suitable circuit to feed your radiators. Usually two circuits are used – one upstairs and one downstairs. If, however, the heating load is high, it may be necessary to incorporate additional circuits.

You should always treat the domestic hot water circuit separately, although the basic method of calculation is the same, with the calorifier in the hot water cylinder being treated as a high output radiator.

If you are installing a solid fuel boiler, the domestic hot water system should be gravity-fed and ideally the bathroom towel rail should be included on that circuit. You can use a gravity-fed circuit with other fuelled boilers, of course. This has the advantage that no electricity for a pump need be consumed during the summer months. A pumped circuit does, however, give you a quicker rate of heating and closer control of the hot water temperature and also allows you to install smaller bore pipes in the circuit.

Provided that the hot water cylinder is positioned higher in the circuit than the boiler, but is still fairly close to it, circulation via gravity presents few problems. The recommended pipe sizes depend on the size of the indirect cylinder as follows: up to 117 litres – 28 mm, 140–162 litres – 35 mm, 190–245 litres – 42 mm.

These pipes should be installed as vertical as possible and there should be no horizontal sections of

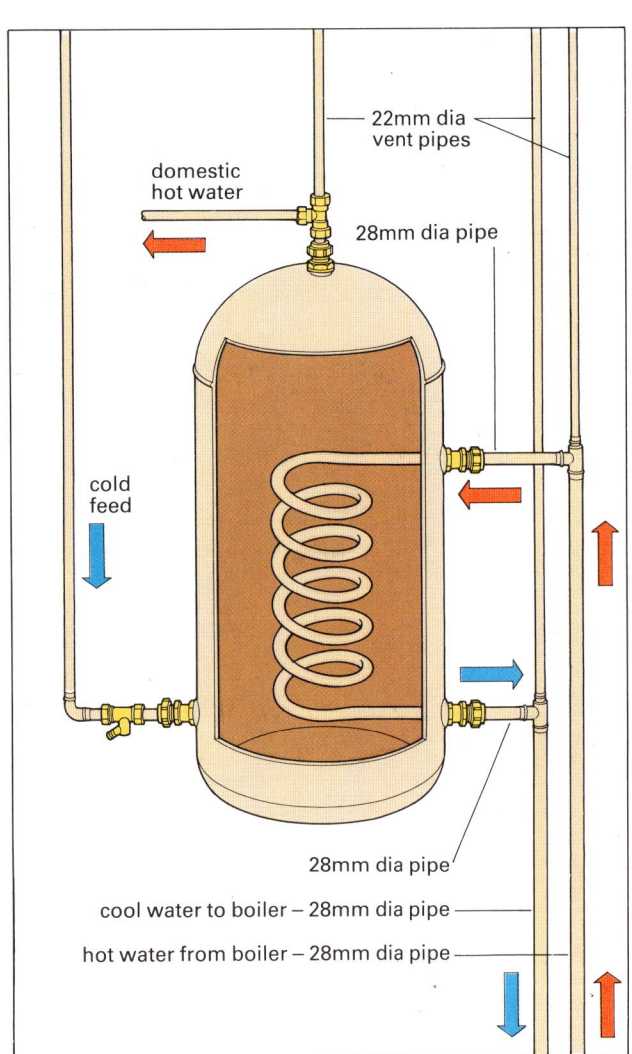

domestic
hot water

22mm dia
vent pipes

28mm dia pipe

cold
feed

28mm dia pipe

cool water to boiler – 28mm dia pipe

hot water from boiler – 28mm dia pipe

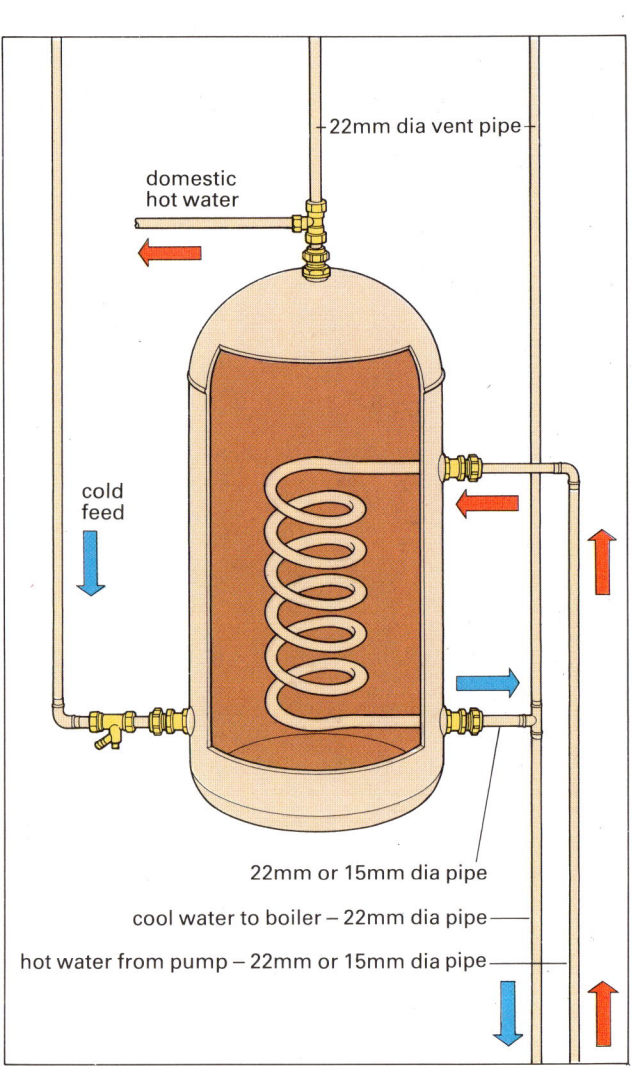

domestic
hot water

22mm dia vent pipe

cold
feed

22mm or 15mm dia pipe

cool water to boiler – 22mm dia pipe

hot water from pump – 22mm or 15mm dia pipe

Here you can see the difference in design of a primary circuit that is gravity-fed (left) and one that is pumped (right). With the gravity-fed system, the circuit relies on convection for the hot water to circulate through the

calorifier in the hot water cylinder. You will need to use larger pipework to ensure less flow resistance. Smaller pipework can be installed with the pumped system, since the hot water is pushed through the calorifier under pressure.

pipe run. If the hot water cylinder is on the same level as the boiler – or a long way away – you will need to seek advice on how to plumb in the system. This particular problem is thoroughly investigated in a booklet published by the Solid Fuel Advisory Service – *The Design of Domestic Central Heating Systems*.

If you are installing a pumped primary system, you can use the previous method of calculation to establish the pipe sizes, taking the calorifier as a radiator.

Calculating the pump size

It is not necessary at this stage to decide exactly where the pump has to be sited. All the information you need is the rate of water flow that the pump is required to move and the resistance it has to face.

The resistance is measured in Newtons/m². In the average house there will be two circuits – the primary one and the secondary, with the former offering little resistance to the flow of water. The total pressure losses for each circuit should be checked; the circuit with the highest loss is known as the 'index' circuit. It is this circuit that determines the pressure developed by the pump. Since the other circuit has a lower resistance, it should overcome this.

The index circuit will usually be the one that carries the highest heating load and is also the longest. If you are fitting convector heaters, these usually determine the index circuit since they have a higher resistance to flow than normal radiators.

The rate of flow is determined by taking the total heat emission requirements for the house – that is, all the radiator emissions, pipe losses and the emission from the domestic hot water. To this total you need to add a further allowance – that of the boiler – to compensate for those days when the outside temperature is below −1°C. Without this, the boiler would be unable to supply sufficient heat to maintain the desired temperature. The allowance usually applied is 15 per cent for solid fuel boilers and 10 per cent for boilers using any of the other fuels.

The total heat emission plus the boiler allowance is the total heat load for the whole house. If the domestic hot water system is gravity-fed, the emission requirement for this should be deducted from the total heat load. If the calorifier circuit is fed via a pump, you should use the total heat load in the calculations.

To work out the equivalent heat flow, take the total heat load and divide this by the temperature drop in the radiators (normally 10°C). Now divide this figure by the specific heat capacity of water (4180 joules/kg°C), which will give the rate of flow in kg/sec.

These calculations can be summarised as follows: total radiator emissions + total pipe emissions + domestic hot water emission = total house emission.

total house emission + boiler allowance = total heat load.

For a pumped primary circuit:
$$\frac{\text{total heat load}}{\text{temperature drop}} = \text{pump equivalent heat flow}$$

For a gravity-fed primary circuit:
$$\frac{\text{total heat load} - \text{domestic hot water emission}}{\text{temperature drop}} = \frac{\text{pump equivalent}}{\text{heat flow}}$$

The final calculation is as follows:
$$\frac{\text{pump equivalent heat flow}}{4180} = \text{pump rate of water flow (kg/sec).}$$

The work load for the pump is the pump rate of water flow against the total pressure loss of the index circuit. Most circulating pumps now have a facility for adjustment so you can match these requirements.

Calculating the boiler size

The boiler size should be such that it is capable of supplying the total heat load as already calculated. It is unlikely that the total heat load you have calculated will match exactly the output of a particular boiler and it is important that you get a boiler with the nearest output above your total heat load.

6 Installing the system

Having calculated the boiler and pump size, the number, size and position of the radiators and also the size of the hot water cylinder, you are now nearly ready to begin the installation. There are, however, a number of variations on the design of the central heating circuit that you should consider before you start the work.

You can see from the diagram a typical house system with gravity-fed domestic hot water and pumped central heating circuits. The larger pipes are used for the gravity-fed circuit. In the other diagram you will see a system for the same house where both domestic hot water and central heating circuits are pumped. This incorporates the use of smaller pipes throughout and enables you to heat up the domestic hot water that much quicker. However you will need to fit additional valves, which will probably offset the cost difference between the sizes of pipe.

You will notice that some features are common to both systems, as follows:

Cold feed pipe   This always comes from the bottom of the feed and expansion tank and is connected to the bottom of the boiler. The pipe should be of a minimum diameter of 20 mm ($\frac{3}{4}$ in) and its route must be as direct as possible. No valves of any kind should be fitted to this pipe, apart from a draincock at the bottom. This is to enable you to drain the system when required for maintenance or repair work. The return pipe from the calorifier or central heating circuit may be teed into this pipe.

Boiler vent pipe   This runs up from the boiler and terminates in a U-bend over the feed and expansion tank. It should be routed in such a way that it rises continuously and never at any stage runs horizontally.

If it has to run along a wall or beneath a floor, you must ensure that there is a slight rise towards the feed and expansion tank.

Once again, no valves should be fitted to this pipe, although it can be used to feed the calorifier. The minimum pipe size you should use is 20 mm ($\frac{3}{4}$ in), although if the circuit is gravity-fed the pipes must be larger. Boiler manufacturers will frequently specify the minimum permissible size. The height of the U-bend above the feed and expansion tank will depend on the pump circulating head (see Pump position below).

Cylinder vent pipe   This runs from the highest point in the hot water cylinder and terminates in a U-bend above the cold water cylinder. No valves should be fitted to it. The domestic hot water supply pipes are teed off from it – at least one to feed the bathroom and kitchen and a separate pipe to feed a shower unit using domestic hot water or a bidet with ascending spray, where either of these are fitted.

Safety valves   Boiler safety valves used to be fitted to all systems, although in some modern systems these may not be required. The information leaflet supplied by the boiler manufacturer will usually tell you if you need to fit such a valve. If it is necessary, you should fit the valve as close as possible to the boiler on the hot water flow pipe. Make sure there are no other valves or tee fittings between this valve and the boiler.

Pump position   The pump is responsible for moving the water round the circuit in which it is fitted. It follows, therefore, that on one side of the pump there is a positive pressure (or head) and on the other side a negative pressure. You can position the pump on either the flow or return pipe to the boiler, but in each case different conditions apply.

In our simplified diagram, the pump is situated on the line of flow to the heating circuit. In this circuit,

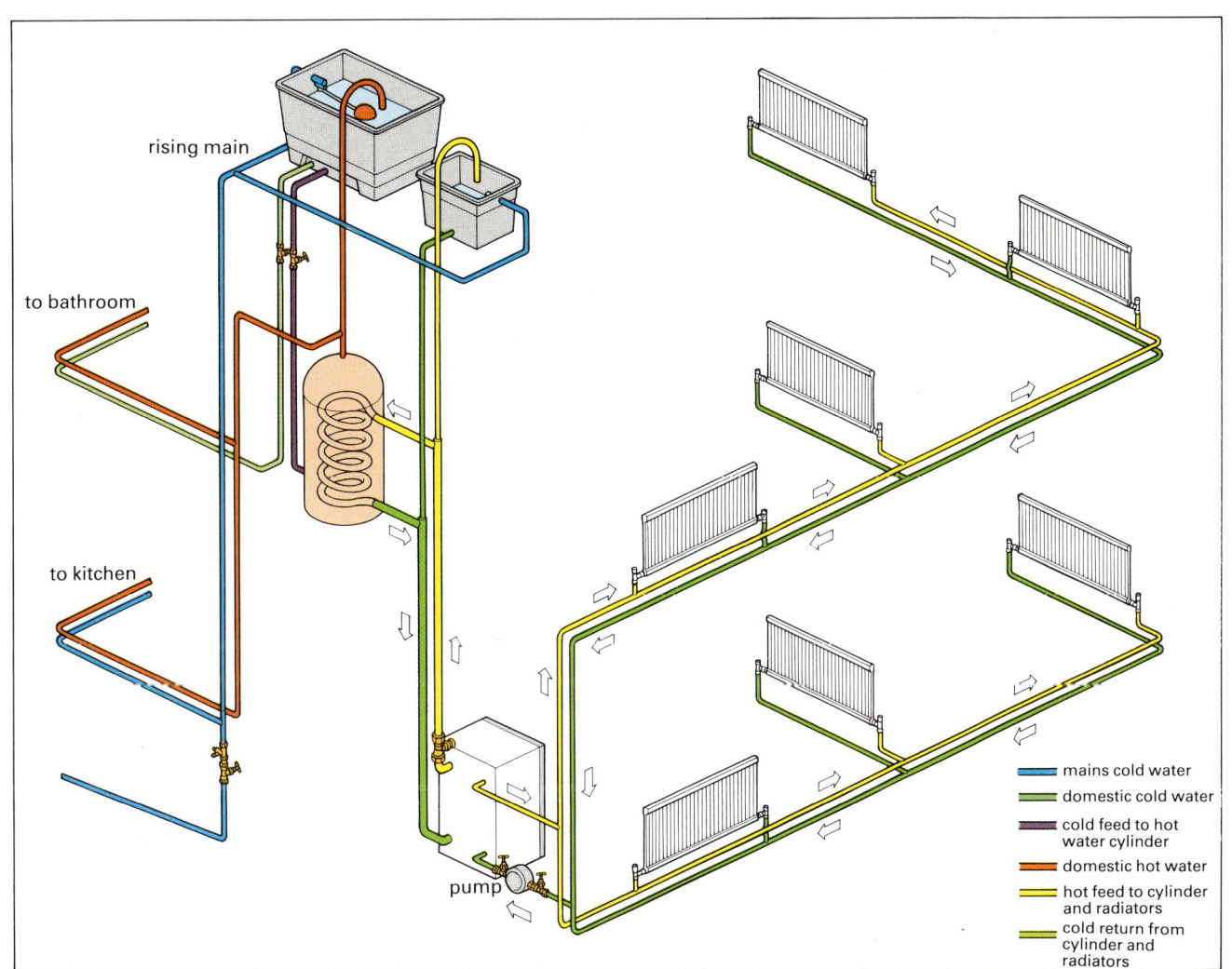

rising main

to bathroom

to kitchen

pump

mains cold water
domestic cold water
cold feed to hot water cylinder
domestic hot water
hot feed to cylinder and radiators
cold return from cylinder and radiators

This typical system incorporates a pumped central heating circuit and a gravity-fed domestic hot water circuit. The heating circuit uses the two-pipe system to feed the radiators and it is normal to have a separate heating circuit for each floor of the house. With larger houses, more than one per floor may be required.

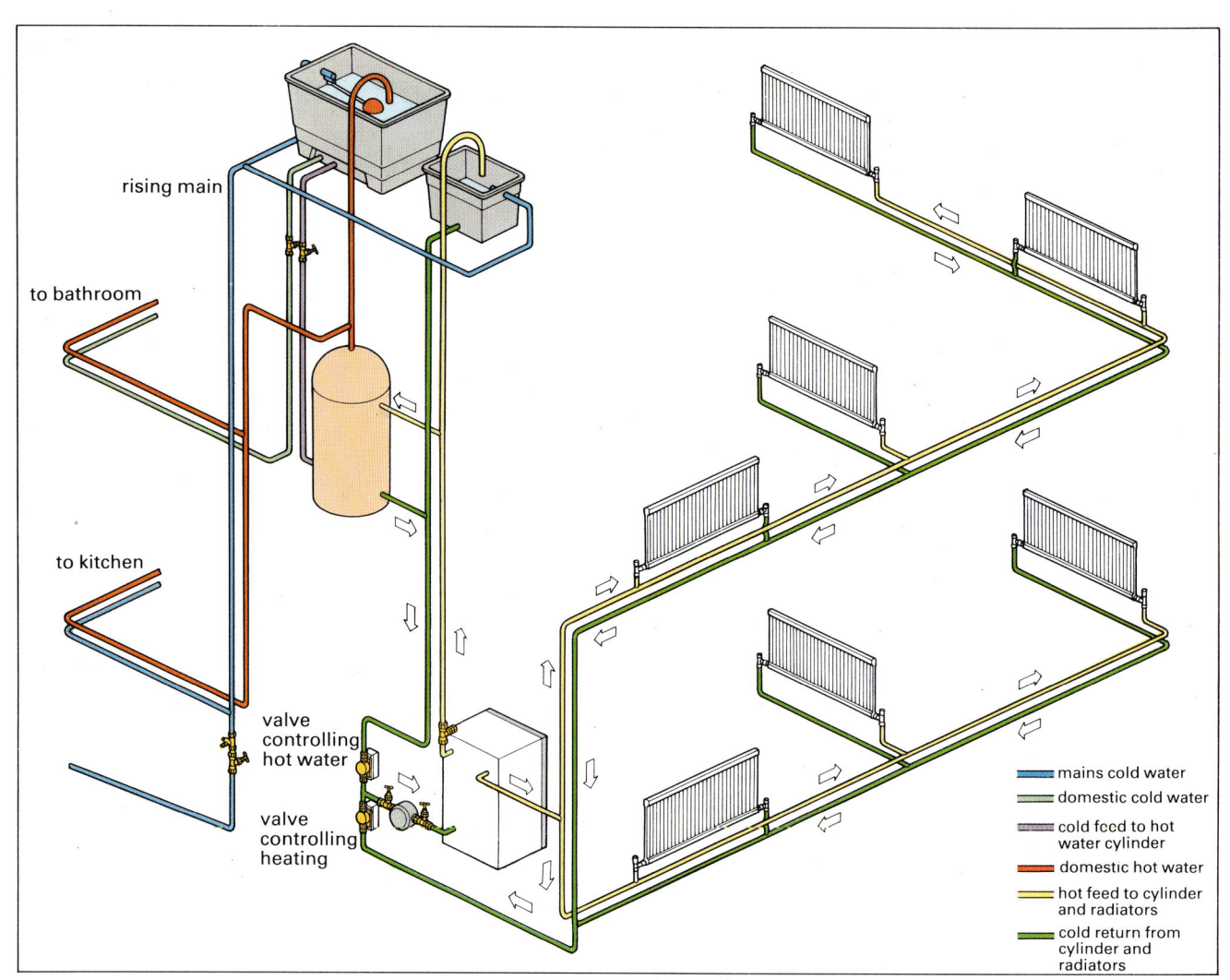

rising main

to bathroom

to kitchen

valve controlling hot water

valve controlling heating

mains cold water
domestic cold water
cold feed to hot water cylinder
domestic hot water
hot feed to cylinder and radiators
cold return from cylinder and radiators

Here you can see a typical system where both the central heating and domestic hot water circuits are pumped. Two motorised valves must be fitted to control the flow of water through each circuit. There has to be a valve on each since the pump will operate whenever the boiler is switched on. When both circuits reach the required temperature, the thermostat on each will shut off the motorised valve. When both valves are shut (boiler and pump) switch off.

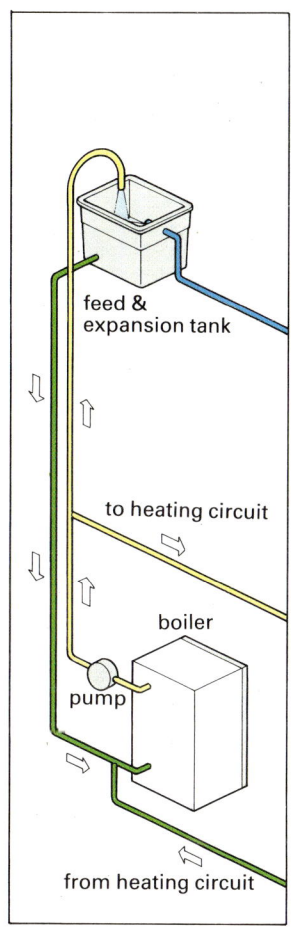

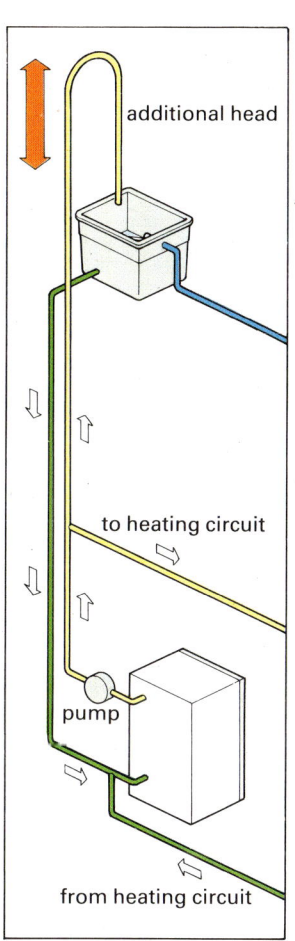

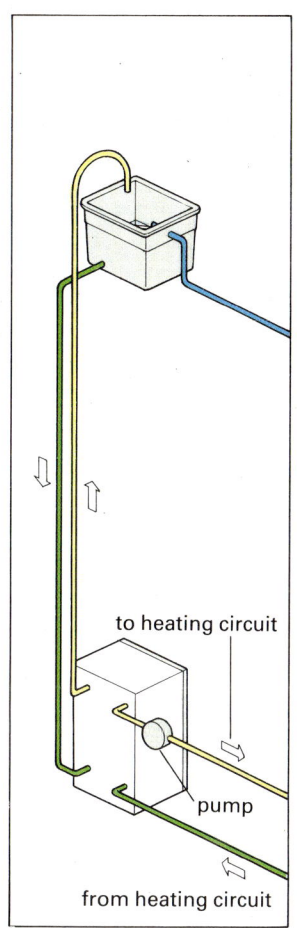

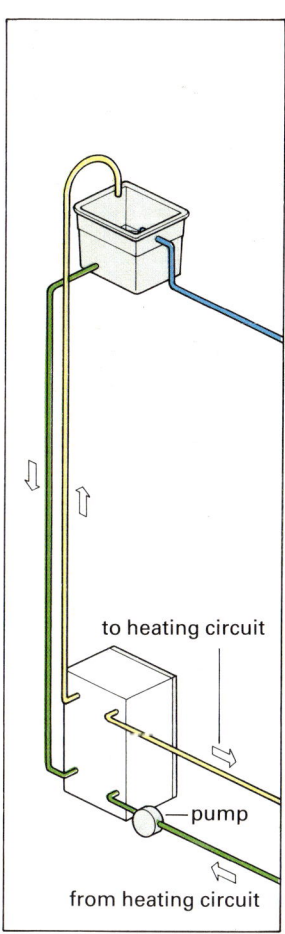

In the diagrams:

feed & expansion tank

to heating circuit

boiler

pump

from heating circuit

additional head

In these four diagrams you will notice different pump positions. In the first case, water could be pumped directly back into the feed and expansion tank if there was too much resistance in the heating circuit. One way of overcoming this is to increase the height of the U bend.

Better still, you can have the feed to the boiler and the vent pipe separate from the pumped circuit. Finally you can draw the water round the system, in which case there will be slightly less pressure.

the water has a choice of two routes when it reaches the tee in the boiler vent pipe. It can either pass along the heating circuit or flow up the vent pipe into the feed and expansion tank.

If the head in the vent pipe is less than the head due to flow resistance in the heating circuit, then the water will flow up the vent pipe. Since the pump will be drawing water from the feed and expansion tank to supply the circuit, the tank will not overflow. But little or no water will find its way round the heating circuit and the net result will be a loft full of steam.

There are three ways in which you can prevent this situation. Fron the diagram, you can see that the height of the U-bend above the feed and expansion tank has been raised to increase the head of the vent pipe. We also show here a different layout where the pump is fitted in the line of flow but the water has to pass around the heating circuit and the vent pipe is not affected by the pump head.

We also show a commonly used alternative, where the pump is fitted into the return line. This means that the central heating circuit is operating at negative pressure. One feature of this situation is that any leaks in the circuit are more likely to draw air in than let water seep out.

Whichever type of circuit you design, you should always ensure you can isolate the pump. To do this you simply install a gate valve on either side, using union fittings. This enables you to remove the pump when required for servicing or when filling the system without having to drain it all off first.

Pump controls   In a simple system where the domestic hot water is gravity-fed, the pump is only used with the central heating circuit. By switching off the pump, you will stop the flow of water in this circuit. This is usually done by means of a suitably positioned room thermostat (see pages 50–51).

The boiler will keep running until the water in the

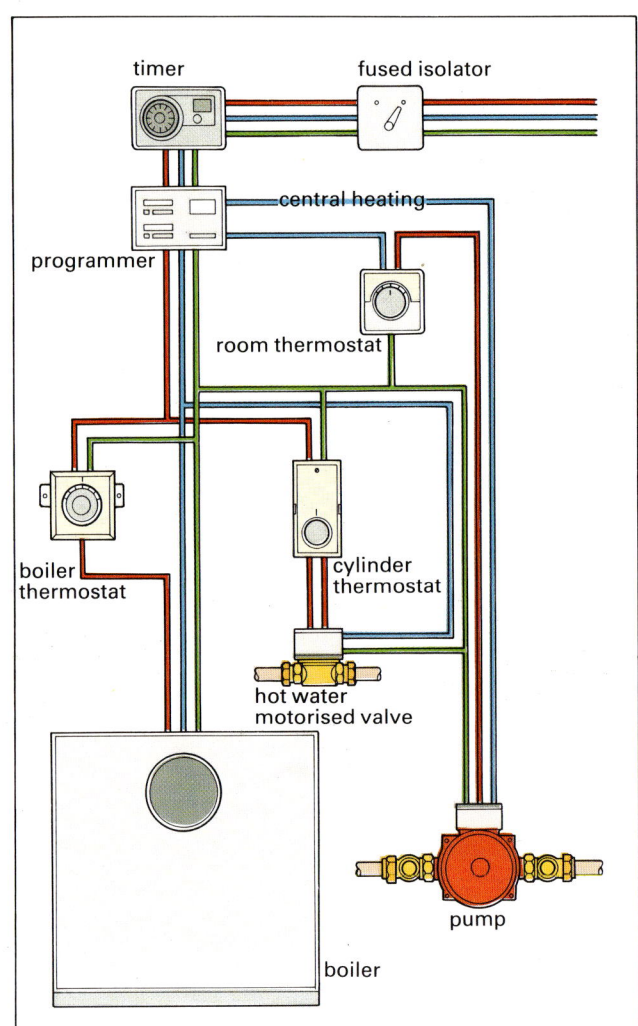

This is a typical control circuit where the domestic hot water is gravity-fed and the central heating pumped. The boiler thermostat controls the calorifier. Individual circuits will vary according to the type of controls used.

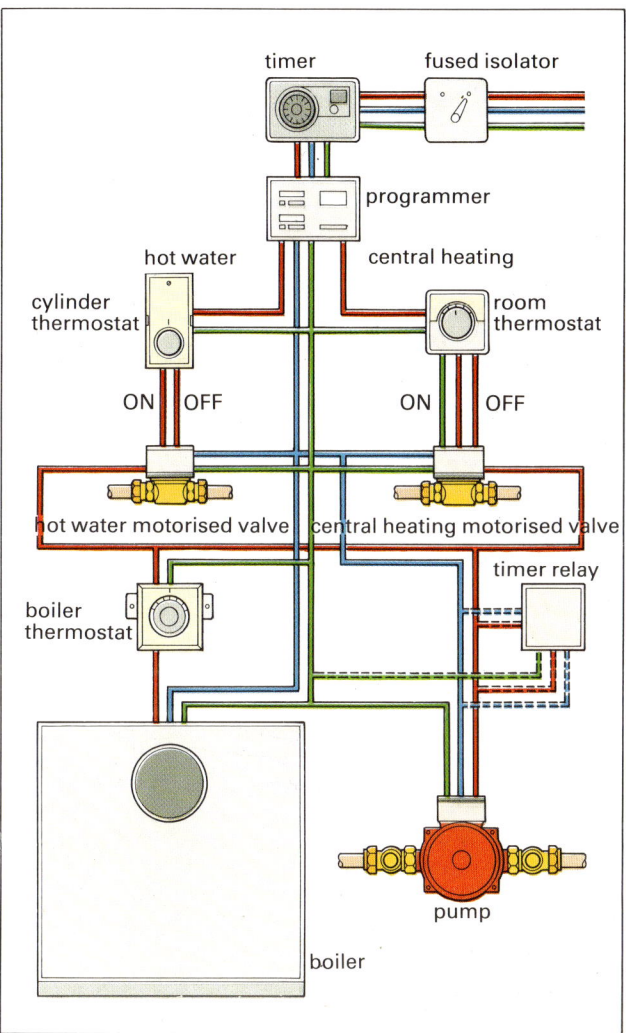

timer

fused isolator

programmer

hot water

central heating

cylinder
thermostat

room
thermostat

ON   OFF

ON   OFF

hot water motorised valve

central heating motorised valve

timer relay

boiler
thermostat

pump

boiler

*This control design is for a fully pumped system. The
timer relay may be required for bypass pipework if a
wall-mounted boiler is used. This circuit is diagramatic
and each circuit will vary depending on the controls used.*

domestic primary circuit reaches the temperature to
which the boiler thermostat has been set. It then
switches off and will remain that way until either the
central heating thermostat calls for heat and switches
on the pump or a hot water tap is turned on
somewhere in the house. Either of these will cause a
drop in the temperature of the water in the boiler and
its thermostat will turn it on again – provided, of
course, that the system's timeswitch is on.

If you fit a thermostat to the hot water cylinder,
this will give you better control of the domestic hot
water temperature since it will switch the boiler off
when the required water temperature is reached.

With a fully pumped circuit, the pump is operating
all the time the boiler is running. Both the domestic
hot water and central heating circuits are controlled by
their own thermostats, which operate two motorised
valves. If either or both of these valves are open, the
boiler and pump will operate. When both circuits have
reached the desired temperature, both thermostats will
close their motorised valves and the boiler and pump
will stop.

Bypass pipework   Some boilers that have a low water
capacity need water to circulate for a time after they
have stopped. In such cases you will need to install a
bypass loop of pipework, incorporating a restricting
valve, through which water can continue to circulate
during this period.

The pump is kept operating by a timing relay
which supplies current to the pump for a short period
after the boiler has stopped. To prevent water flowing
through the bypass loop when the system is working
normally, the restricting valve is adjusted to give a
higher resistance to flow than the index circuit or the
primary circuit.

Check with the information leaflet supplied with
the boiler to see whether you need to incorporate any
bypass pipework in your system.

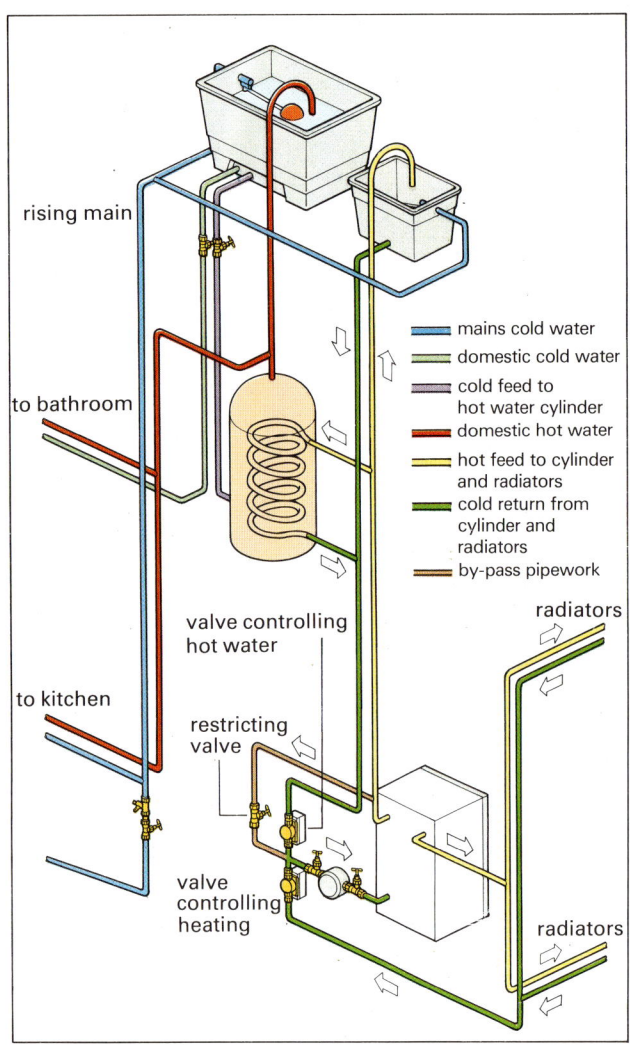

rising main

to bathroom

to kitchen

valve controlling
hot water

restricting
valve

valve
controlling
heating

radiators

radiators

mains cold water
domestic cold water
cold feed to
hot water cylinder
domestic hot water
hot feed to cylinder
and radiators
cold return from
cylinder and
radiators
by-pass pipework

Bypass pipework as shown here is sometimes required with wall-mounted or other boilers that have a low water capacity. This is to allow circulation after the boiler has switched off to prevent the water boiling inside.

The installation work

For the purposes of this section, we have assumed that you are installing a complete new system. It is equally possible to convert an existing direct hot water cylinder by inserting a calorifier through one hole drilled at the top and feeding it out through another hole at the bottom of the cylinder. Alternatively, in the case of a pumped primary circuit, you can fit a special heating coil into the immersion heater boss.

If you are considering either of these alternatives, you must first make sure that your existing cylinder is the correct size, in good condition and free from scale. You probably will not find it particularly easy to remove your existing immersion heater due to corrosion in its large diameter thread. You will almost certainly need the help of a special immersion heater spanner, which you may be able to hire, and some penetrating oil.

If you do want to fit a hot tube calorifier in the immersion heater boss, you must bear in mind that this will replace the existing electric immersion heater. By fitting the coil-type calorifier inside the cylinder, you can retain your immersion heater and therefore keep the facility to heat the domestic hot water in emergency – or, of course, during the summer without switching on the boiler. If you are installing the coil-type, the manufacturer will supply full instructions and details of the special fittings required.

In order to minimise the disruption to the household, which will be inevitable during the period of installation of the central heating system, you should consider tackling the work in the following order:

Fitting radiators Normally you should fit these horizontally. Where the wall space is limited and this proves impossible, you can fit them vertically instead, although this will not give you such a wide spread of heat from them.

Radiators are normally positioned so that the

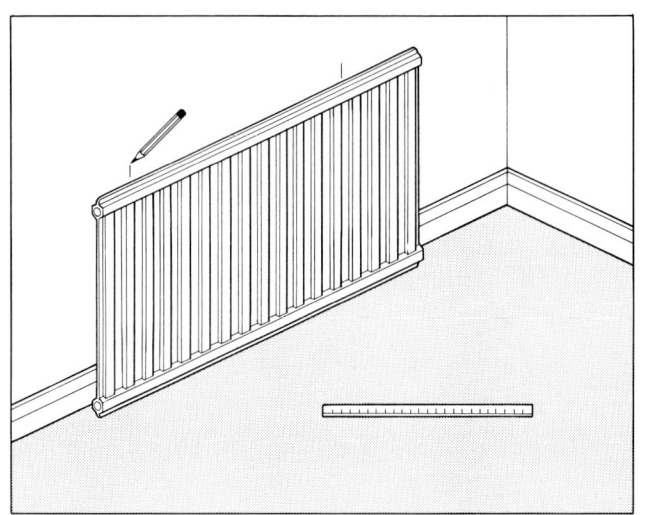

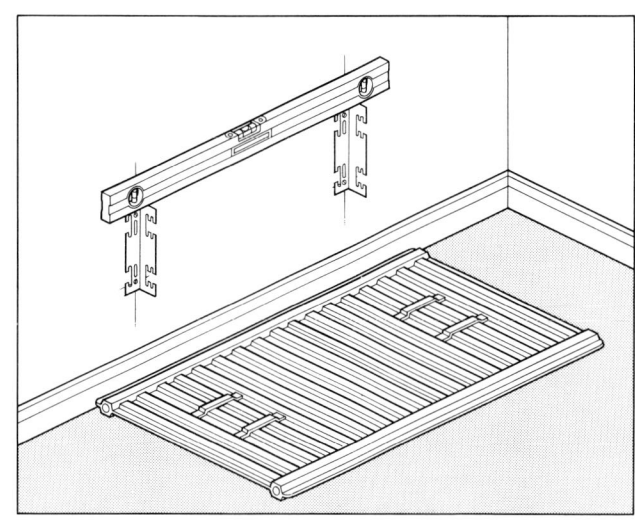

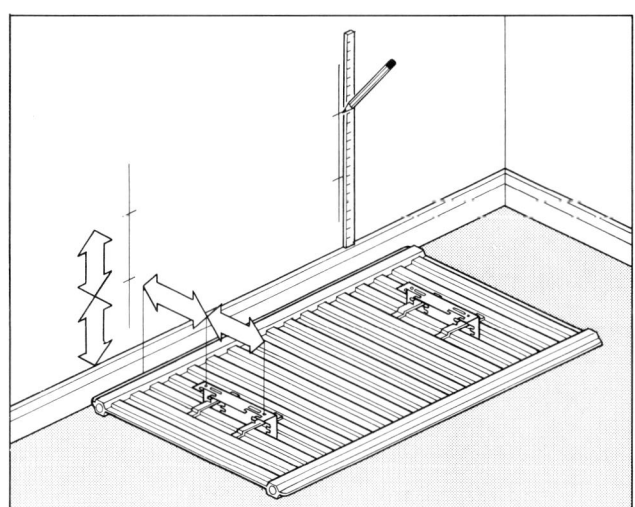

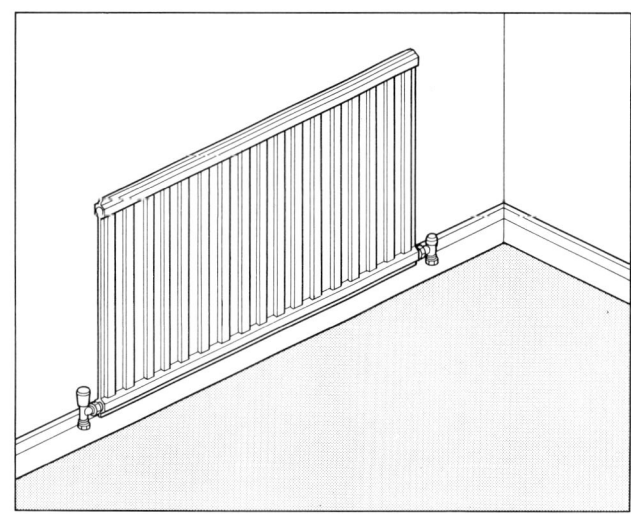

When hanging a radiator on to a wall, first mark the vertical position of the brackets (top left). Then mark the correct height for the brackets (above left) so that the radiator is in the right place when hung. Fix the brackets

with 35 mm (1½ in) screws, checking with a spirit level to ensure the radiator will hang straight (top right). Fit on the relevant controls – the radiator and lockshield valves – and hang the radiator in position (above right).

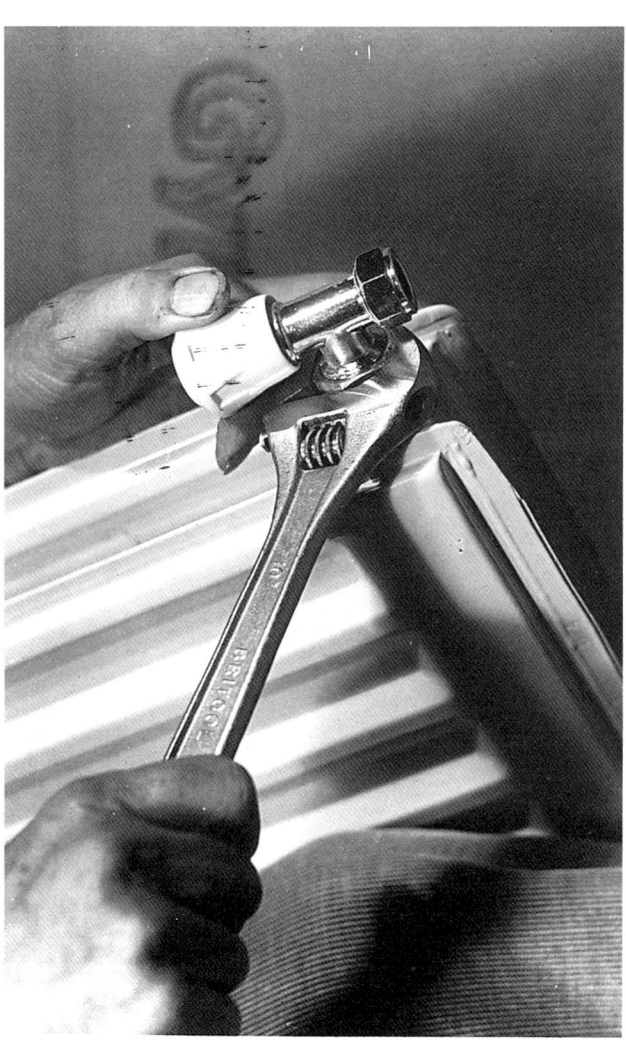

Since the radiator brackets have to carry a fair amount of weight, make sure they are firmly fixed to the wall. This means drilling and plugging suitably sized holes to take 35 mm (1½ in) screws.

You will find it a lot easier to fit the relevant control valves on to the radiator before you hang it in place on the wall. To avoid possible damage to the fittings, always use a good spanner and not a pipe wrench.

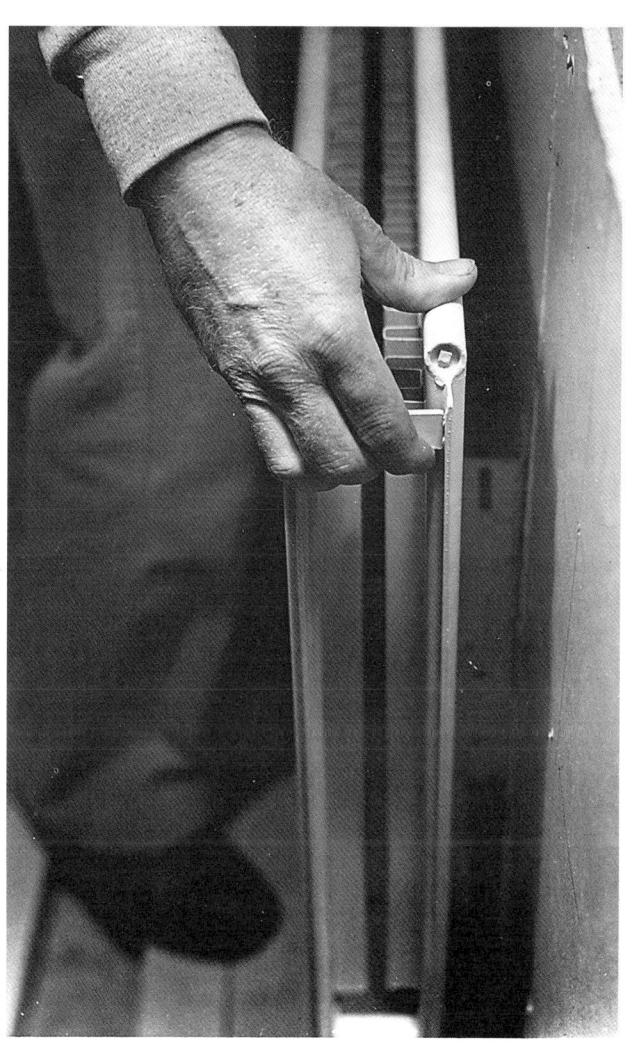

Before you lower the radiator in place, check that you know where the straps on the back fit. They should drop into the slots in the brackets. Since the radiator is heavy, you do not want to keep lifting it up to the wall.

Although you should be able to manage the average size radiator on your own, if in doubt it is always best to seek an extra pair of hands to lower it into position. The air-release valve should be in position in the top.

bottom is level with the skirting board. One point to bear in mind is that you will find it a lot easier to decorate the wall behind before you hang each radiator on its support brackets. You may also want to fit heat-reflecting panels on this section of the wall as well. When measuring up, you should also make sure where a radiator is positioned under a window that there is sufficient clearance to enable the curtains to hang behind it. If not, you will have to mount the support brackets on blocks of wood. You may have to shorten the curtains so that they clear the support brackets.

Having hooked the support brackets on the back of the radiator, rest it on the floor against its final position on the wall (see diagram). Mark with a pencil vertical lines on the wall to correspond with the centres of the bracket fixing holes. Then measure the distance between the bracket fixing holes and the bottom of the radiator and mark this distance on the wall, measuring up from the top of the skirting board.

Remove the support brackets and offer each in turn to the wall, marking through the fixing holes with a pencil. Using a masonry drill bit on an electric drill, make holes in the wall to take No 12 or 14 wall plugs, since a radiator full of water is very heavy. Using screws at least 35 mm (1½ in) long, fix each bracket firmly on to the wall, checking the second bracket is aligned with the first with the help of a spirit level. This is to ensure the radiator is level when hung on the brackets.

Before hanging the radiator in place, wrap two or three turns of PTFE tape around the tail of each of the two valves to be fitted on the radiator – one lockshield and one shut-off or thermostatic valve. Then screw each tail firmly into place in the two bottom tappings. Fit the valves loosely to the tails, tightening the union nuts by hand only at this stage. You can now hang the radiator on its brackets.

Fitting skirting radiators   First remove the old skirting board (see diagram) and clean up any damage caused to the surrounding plaster. Now offer up the radiator backplate into position and mark out and drill holes for the wall plugs and fixing screws. Offer up the heating element and mark the positions of the holes in the floor for the flow and return pipes, making sure these are not above any joists.

Remove the element and backplate and drill the required holes. If the flow and return pipes are at either end of the radiator, the holes must be drilled about 6 mm (¼ in) larger than the pipe diameter to allow for expansion. This is not necessary when these pipes are fitted to the same end of the radiator since the expansion will be contained within the casing.

Now fit the backplate in position using 32 mm (1¼ in) No 10 screws, with the corresponding size of wall plugs. Put on the end fittings or valves necessary and clip the heating elements into place. You should fix on the front cover to protect the fins until you are ready to plumb in the radiator.

Fitting the boiler   This is probably the most demanding operation and one that must be carried out very carefully. Read the installation instructions supplied by the boiler manufacturer and work to them exactly. You must pay particular attention to any fireproofing requirements in respect of the immediate surroundings and the floor beneath if it is a free-standing model.

If you are installing a balanced flue boiler, check carefully on the instructions for fitting the flue through the wall, bearing in mind the restrictions on the siting of the flue and general access around the boiler for servicing and maintenance (see pages 33–41). Locate the nearest connection point for gas and electricity (if required) and bear in mind that the gas supply piping and connections should be made by a competent engineer, who should also commission the boiler for

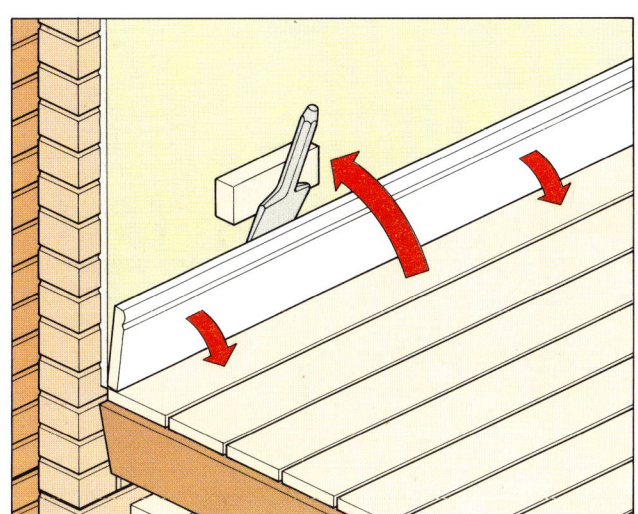

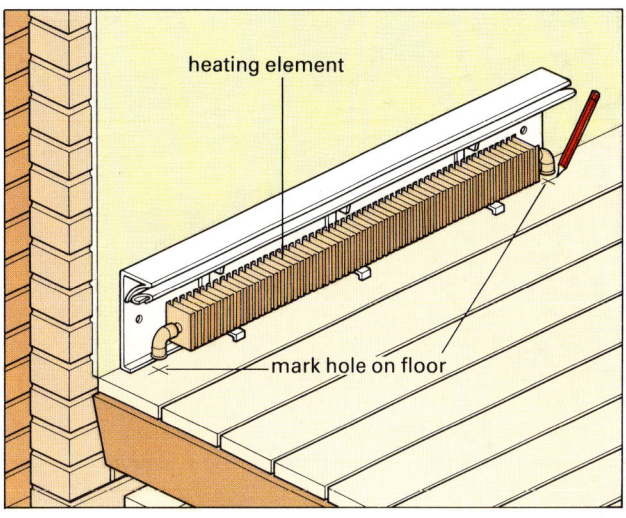

heating element

mark hole on floor

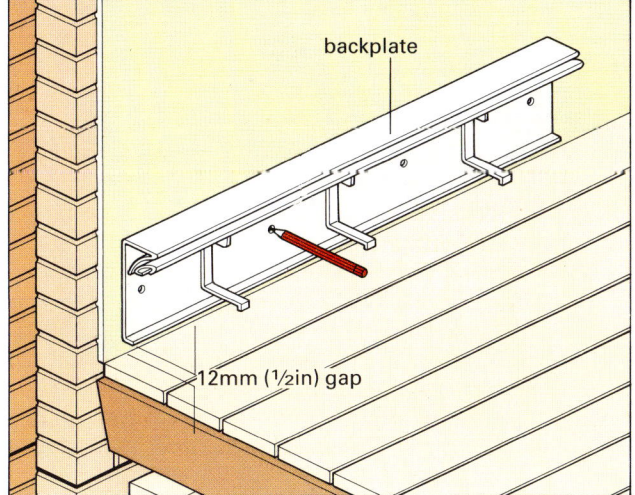

backplate

12mm (½in) gap

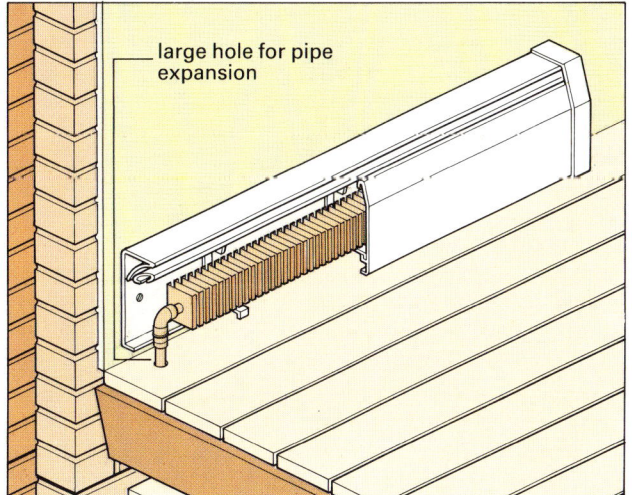

large hole for pipe expansion

When fitting a skirting radiator, prise off the skirting board, packing the chisel to prevent damaging the wall (top left). Holding the backplate in position, mark the fixing holes (above left) then drill and plug these and

screw the backplate in place. Hold the heater element in position and carefully mark the holes for the pipework in the floor (top right). Drill a hole either end slightly larger than the pipe to allow expansion and connect up (above right).

you (see pages 108–111). The same applies to the connection of electricity.

Since the exact installation instructions will vary, depending on the type of boiler you buy, it is impossible to make any general comments. Always check carefully with the manufacturer's instructions and, if in any doubt, seek advice from the manufacturer, supplier or competent engineer.

Fitting the hot water cylinder   The normal location for this is in the airing cupboard. Make sure there is enough room around the cylinder for you to fit an insulation jacket at least 75 mm (3 in) thick, unless you are installing a pre-insulated cylinder. There should still be space around the cylinder to enable air to circulate and ensure an efficient airing process and to provide access to all the plumbing connections, should you need to carry out any repair work later.

Make sure that the base on which the cylinder stands is level and strong enough to take the weight of the cylinder when it is full of water. As an example, a 136 litre (30 gallon) cylinder will contain in excess of 120 kg (300 lbs) of water. The cylinder itself is free-standing, although it will of course be held rigid once the plumbing connections are made.

Fitting the cold water storage cistern   This is normally sited in the loft and, because of its weight, usually near to or above a load-bearing wall. The base on which it stands must be very strong and you should make sure that there is no loft insulation directly underneath it. This is to allow warm air from below preventing the cistern freezing up in extreme weather.

Where you are installing a new cistern, you should buy one made of plastic or glass fibre, since these materials are corrosion-free. Where you have an upstairs bathroom and want to install a shower or bidet, it is a sensible precaution to install the cistern on a raised platform. This ensures you have a better head to maximise the pressure of the water into the fitting.

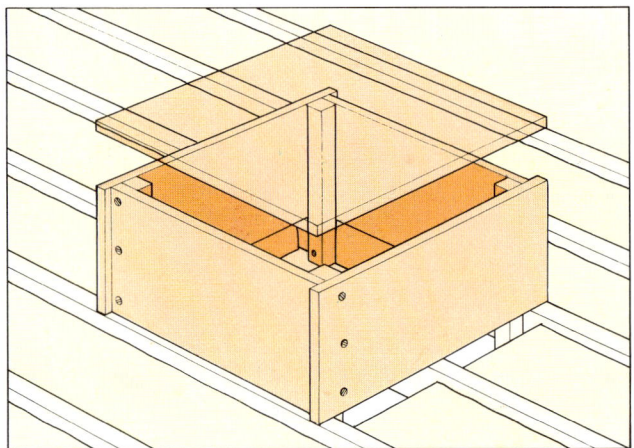

When raising the cold water storage cistern, you must use blockboard and not chipboard panels to ensure the platform is strong enough to hold the considerable weight of the cistern.

Fitting the feed and expansion tank   This is usually installed next to the cold water storage cistern in the loft to minimise the pipe runs. When fitting the tank, you should make sure that the bottom of it is level with or slightly below that of the storage cistern. The reason for this is that should a leak develop in the calorifier due to corrosion, for example, water will then pass from the domestic circuit into the primary circuit. Eventually the leak will be brought to your notice since the level of water in the feed and expansion tank will rise and finally flow down the warning pipe.

If the position of the two were reversed, water would flow from the primary circuit into the domestic circuit and any additives present in the primary circuit would contaminate the domestic hot water. Since the leak would probably only be slight, the continual drawing off of water from the cold water storage cistern would delay or obscure the rise in the water level – and the leak would probably go unnoticed.

Here you can see how the cold water storage cistern has been raised in order to ensure a better head for a shower. When installing a feed and expansion tank, ensure that the water level is always lower than that in the cistern.

Incidentally, the normal size for the feed and expansion tank is 45.5 litres (9 gallons) for systems up to 22 kW.

Lifting floorboards

Unless you are installing the central heating system into a house that is in the process of being built, in which case it is easier to do this through the ceiling before you put that in place, you will almost certainly have to lift some of the floorboards to lay the pipework.

Pipes will have to be fed through floorboards as well and here you will normally have to drill holes that are at least 6 mm ($\frac{1}{4}$ in) larger than the external diameter of the pipe you are using. Not only will this allow for any minor errors in alignment, but will also take into account any expansion and movement in the pipes.

Before you lift floorboards, examine the floor closely to see whether any of the boards have been lifted before. New nails, clean screws or marks on the edges of boards will be the obvious clues. The chances are that certain floorboards may have been lifted before to get at plumbing pipework or electrical cables running underneath. As a general rule, it is a good idea to screw lifted boards back in place – and this will provide the best clue.

Before you attempt to lift any floorboards, check to see whether they are tongued and grooved. To do this, slide a thin blade of a knife either side of the board to be lifted. If the blade hits an obstruction, then the board is tongued and you will have to cut through this to release it. Cut through on either side of the board with a tenon saw held at a shallow angle. This should prevent you cutting through any existing cable or piping underneath.

Start lifting from a convenient end by inserting a bolster or wide-bladed cold chisel and levering up the board. When there is room, slip a long cold chisel or

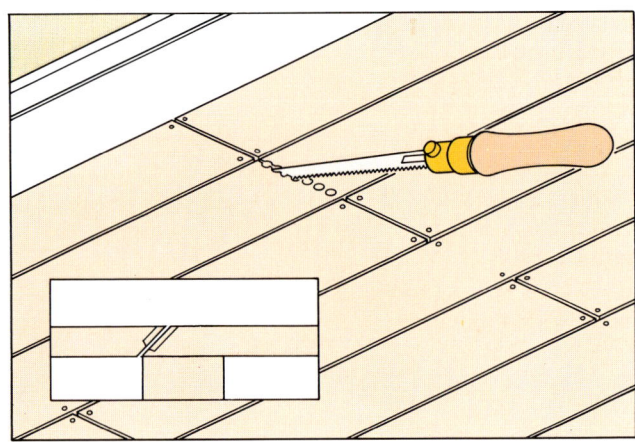

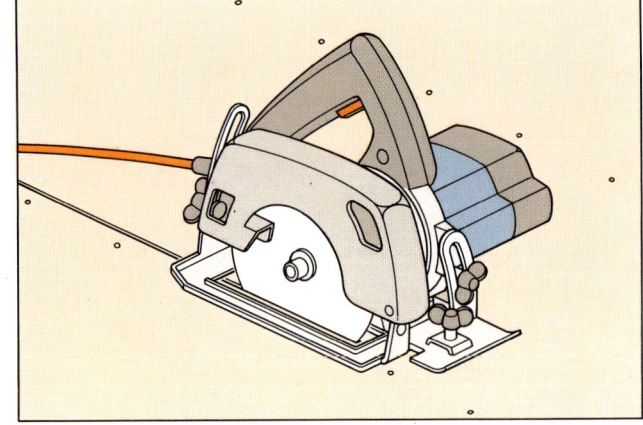

When you have to lift floorboards, first check to see whether the boards are tongued and grooved. If so, you must cut through the tongue to release the first board. Saw at a shallow angle (top left) to avoid any pipes or cables underneath. Having levered the end of the first board up with a bolster chisel, use a steel rod to prise it loose (above left), moving this slong the board's length and tapping it with a hammer if necessary. If you have

to cut across the middle of a board, first drill a series of holes at an angle so that you can get in the blade of a padsaw. Then cut across the board at an angle (top right) so that you can fix through both parts of the board when you lay it back down. If you have flooring panels, these are best cut with a circular power saw (above right), set to the correct depth, above one of the joists to which they are fixed.

steel rod under the board and tap this along the board, lifting it up as you go. You may have to use a lot of brute force at the other end of the board if this is under a skirting board. But be careful not to damage the skirting, which may otherwise have to be removed.

If there is not a convenient end of board at which to start lifting, you will have to cut across the board in the centre of a joist (indicated by the position of the nails or screws). First remove any screws or punch down nails well below the surface. Drill a few small holes at an angle so that you can insert a pad saw to start the cut across the board. Once you have cut right across, you can proceed to lift the board as before. The reason you should cut through the board at an angle is that when you replace the board you can fix through both angled sections of the cut.

When replacing the boards after fitting your pipe runs, fix them into place with 50 mm (2 in) long No 8 or No 10 countersunk screws. This will make lifting much easier if you have to work on that part of the circuit again.

Having replaced lifted boards, you may find there is a gap between the boards. You can make good – and prevent draughts – by using a non-setting mastic filler.

Notching joists

When you are running pipes under wooden floors, you should where possible arrange this so that the pipes run parallel to the joists. Where you have to run the pipes across joists, you should do this by cutting a small notch or groove across the joist to take the pipes.

It is very important that the cut is not too deep since this could weaken the joist and affect the safety of the structure. Pipes are normally run in pairs and this means that you should, where possible, make two small cuts rather than one deep one. Avoid sharp corners at the base of each cut by drilling two 12 mm ($\frac{1}{2}$ in) holes side by side and cut away the waste (see

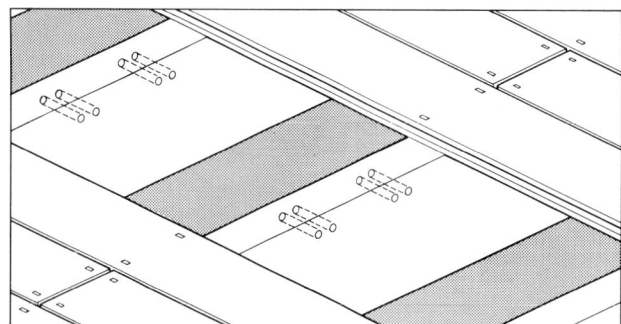
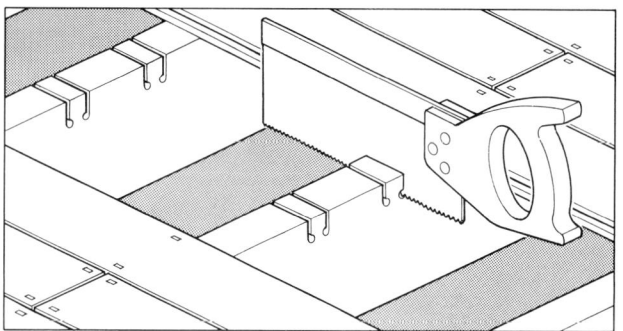
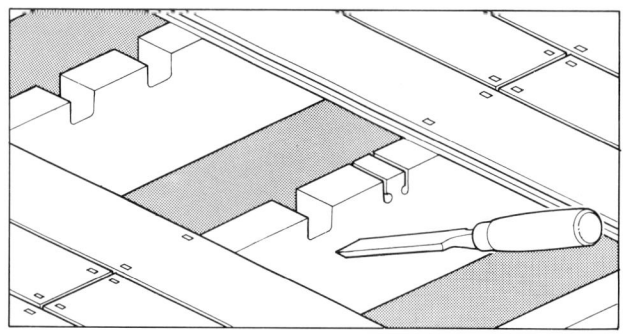

This is the sequence for cutting notches in joists, through which to run pipework. First drill two holes to the width of notch required, then saw down to the outside edge of each and chisel out the waste with the bevel side down.

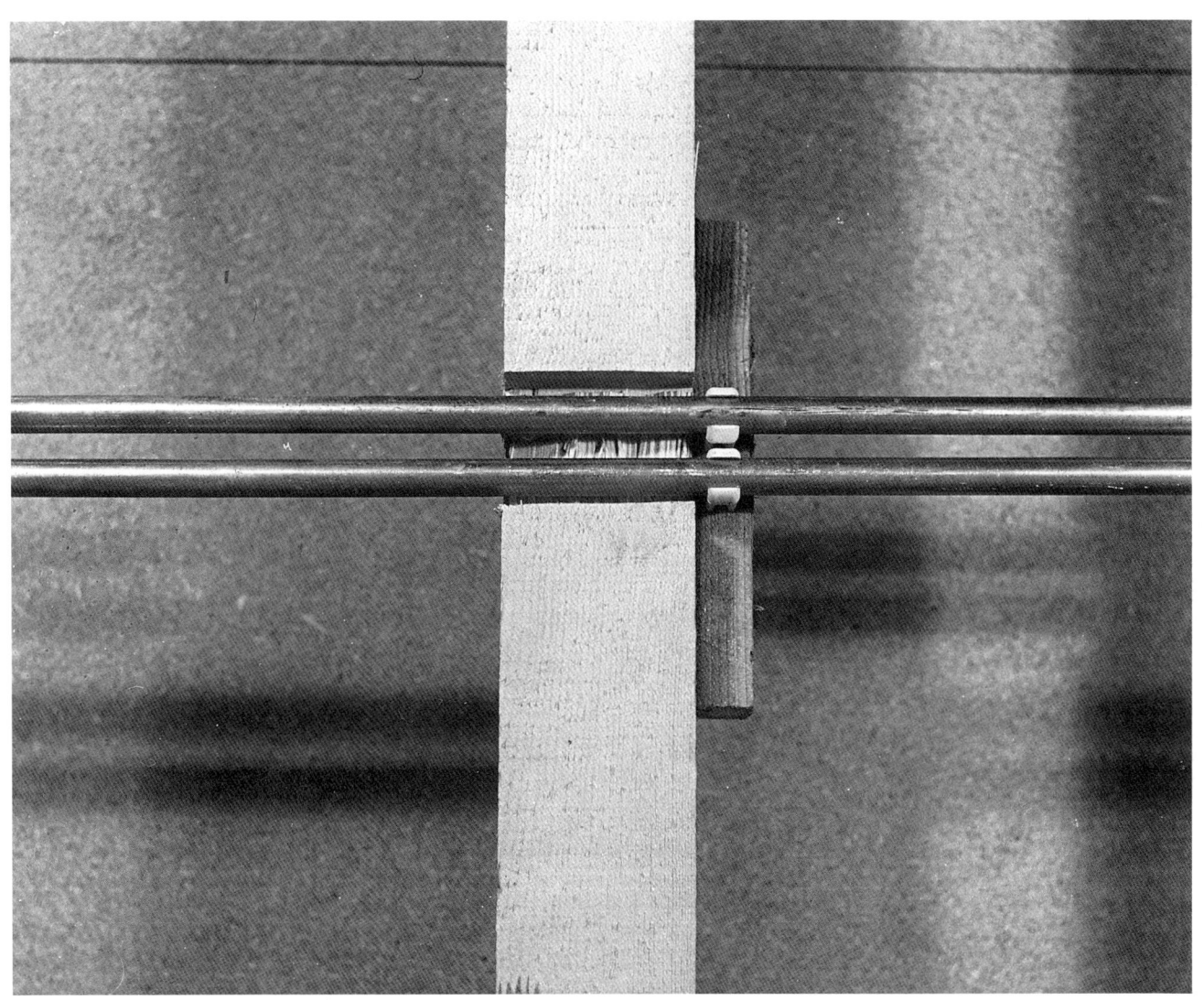

When notching joists, you must keep the notches as small as possible to prevent weakening the timber. However you must allow a little extra space for expansion. Here you can see the pipes running through a notch and clipped in place.

diagram). Make sure that the pipe fits loosely in the cut to allow for expansion. To ensure that any pipe movement is silent, it is best to lay the pipe on a small piece of polythene or nylon.

Plumbing in

It is impossible to give specific details here as to how to plumb in a system, since each one will vary to some degree. Certain features will, however, be common to all systems and these are detailed below, working from the loft downwards.

Cold water storage cistern　You will have to drill a minimum of four holes in this cistern. A control valve, normally a Garston or Torbeck, must be fitted near the top where the rising main enters the cistern. A warning pipe, which runs across the loft to the outside of the house, needs to be fitted about 50 mm (2 in) from the top. This pipe should slope continuously downwards and emerge just below the soffit under the tiles in such a position that any discharge will be easily noticed.

Approximately 25 mm (1 in) from the bottom of the cistern you will need two holes to take the service pipes – one to supply the cold water outlets and the other to supply the hot water system. Special fittings are available which fix into the side of the cistern and are sealed with rubber or plastic gaskets. The pipes from these holes run down below the ceiling, where you should fit gate valves. These enable you to isolate the water supply and carry out any maintenance or repair work without having to drain off the water from the cistern.

Bear in mind that if you have a shower unit or ascending spray bidet fitted, you will need a separate feed from the cold water storage cistern.

The other pipework to the cistern will be the vent pipe from the hot water cylinder. This terminates in a U-bend over the cistern. Make sure that the end of this

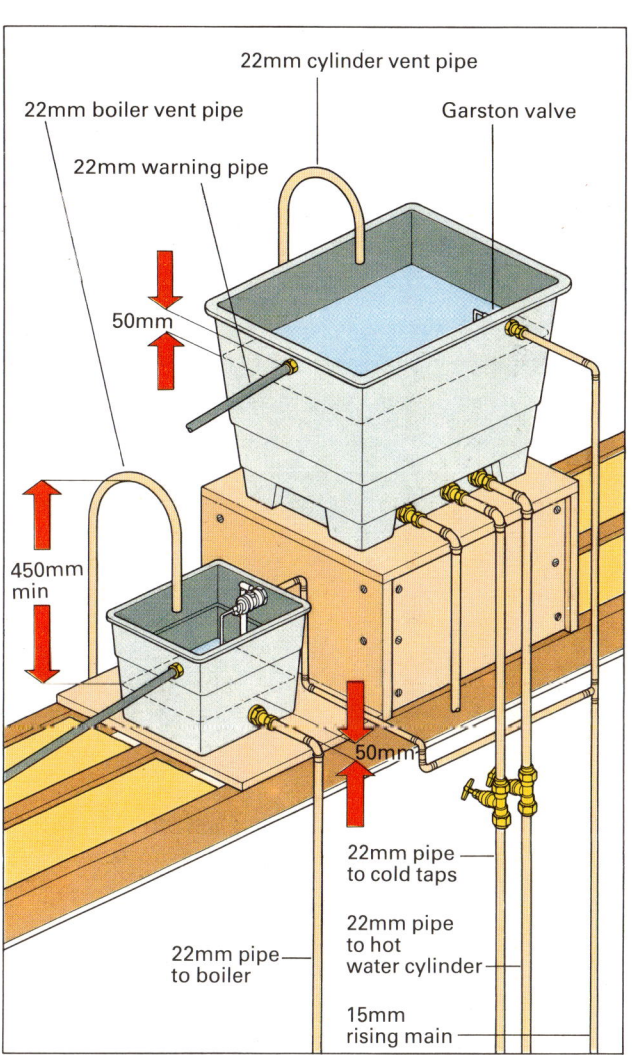

22mm cylinder vent pipe

22mm boiler vent pipe

Garston valve

22mm warning pipe

50mm

450mm min

50mm

22mm pipe to cold taps

22mm pipe to hot water cylinder

22mm pipe to boiler

15mm rising main

This diagram shows a typical arrangement in the loft, where a central heating system has been installed. The third feed pipe from the bottom of the cold water storage cistern is the supply pipe to a shower.

pipe is below the top of the cistern, but not below the level of the warning pipe. If it is and the level in the cistern rises, cold water could get into the vent pipe and be siphoned back into the hot water cylinder when a hot tap is turned on.

When filling the cistern, the control valve must be adjusted so that the water supply from the rising main shuts off when the level is approximately 50 mm (2 in) below the level of the warning pipe.

Feed and expansion tank As with the cold water storage cistern, a control valve with a float suitable for hot water must be fitted near the top of the tank. The service pipe to this valve should be taken from the rising main. To make any maintenance work on this and the cistern control valve as easy as possible, you may like to fit a stopcock just below this junction in the rising main.

You will also need to fit a warning pipe to the tank. Do this in a similar way and position to that on the cold water storage cistern. These warning pipes must be kept separate, so that that if one is overflowing you know where the problem lies.

The feed pipe to the boiler should be taken either from the bottom of this tank or 25 mm (1 in) up the side.

The vent pipe from the boiler terminates in a U bend above the tank. Remember that there must be a mimimum of 450 mm (18 in) vertical distance between the water level in the tank and the top of the U bend. In some circumstances you may need to allow a greater distance (see Pump position above).

When filling the feed and expansion tank and primary circuit, you will need to adjust the control valve so that it shuts off the water supply when the level inside the tank reaches about 50 mm (2 in) above the feed pipe connection. This is to allow for the expansion of water in the circuit when it is hot.

Hot water cylinder With a normal cylinder the pipework connections are quite straightforward. The vent pipe from the cylinder is connected at the top, while the feed from the cold water storage cistern is connected to the single outlet at the bottom of the cylinder. There are two connections for the calorifier, one vertically above the other. The feed to the calorifier is connected to the higher of these, while the return pipe to the bottom of the boiler is connected to the lower one.

Depending on the details of your particular circuit, these two connections may be teed from the boiler vent pipe and the boiler cold feed pipe respectively, if necessary.

Boiler The pipework connections here will vary according to the system. However the cold feed will always enter at the bottom of the boiler and a draincock should normally be fitted near to the connection since this is the lowest point in the system.

The boiler vent pipe (for the primary central heating circuit) will be connected to the top of the boiler and here a safety valve should be fitted if recommended by the manufacturer.

With many boilers you will find similar tappings on both sides. It is usual in these cases to use one side for the feed and vent pipes (combined with a gravity-fed domestic circuit) and the other side for the central heating circuit (see diagram).

The pump pipework will also normally fit into the boiler casing (see diagram for example). Remember to fit a gate valve on each side of the pump, using union connectors so that you can remove the pump easily before filling the system and if maintenance work is required later. You should at this point make up a short length of pipework with suitable fittings (see page 22) to replace the pump when the system is being filled.

If you need to fit a pump bypass (see page 87), you

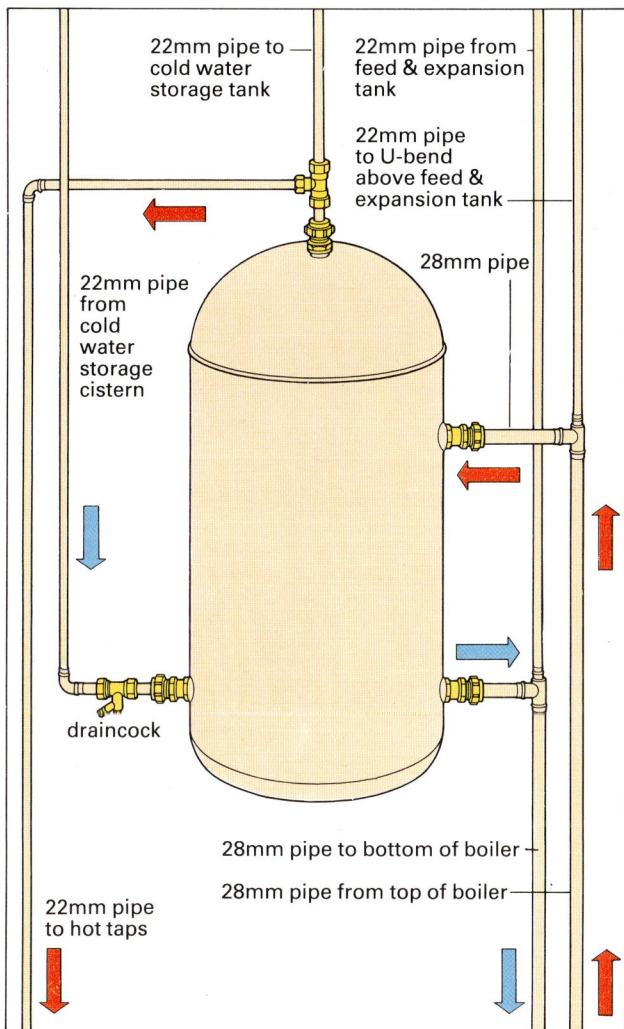

22mm pipe to cold water storage tank

22mm pipe from feed & expansion tank

22mm pipe to U-bend above feed & expansion tank

28mm pipe

22mm pipe from cold water storage cistern

draincock

28mm pipe to bottom of boiler

28mm pipe from top of boiler

22mm pipe to hot taps

Here you can see a typical pipework arrangement for a hot water storage cylinder. The actual size of the flow and return pipes may vary depending on the size of the cylinder and the type of circuit being installed.

should normally be able to incorporate this into the pump casing.

Radiators Because of the possible variations in systems and circuits, there is no way that specific information can be given here. However, there are certain points to bear in mind in terms of general guidance when plumbing in a radiator.

Make sure that all connections to a radiator are sound and that all pipework runs straight out of each fitting. Runs of pipe must always be supported – to walls or joists – with proper plastic pipe clips at correct intervals (see Table G). Where pipe passes through a wall, you should ensure the opening through the wall is lined with a suitable size piece of plastic piping to prevent the expansion of the copper pipe from causing any damage.

When planning the pipework, try to avoid the use of elbow fittings wherever possible: bends in pipework offer far less resistance to the flow of water. Also try to ensure that all pipes are fitted so that no loops or high spots occur which could result in airlocks (see diagram). If a vertical loop is unavoidable, make sure you fit an air-release valve at the highest point on the loop.

Once you have connected up all the pipework in the central heating system, you can lay back those floorboards that had to be lifted. But do not fix them in place until you have checked out the complete system for any possible leaks. You should also leave any lagging of the new pipework until after the check has been made.

Microbore system

As the name implies, this type of system – also known as mini bore – uses smaller pipes than the small bore system. The pipes used are 6, 8, 10 and 12 mm external diameter (OD).

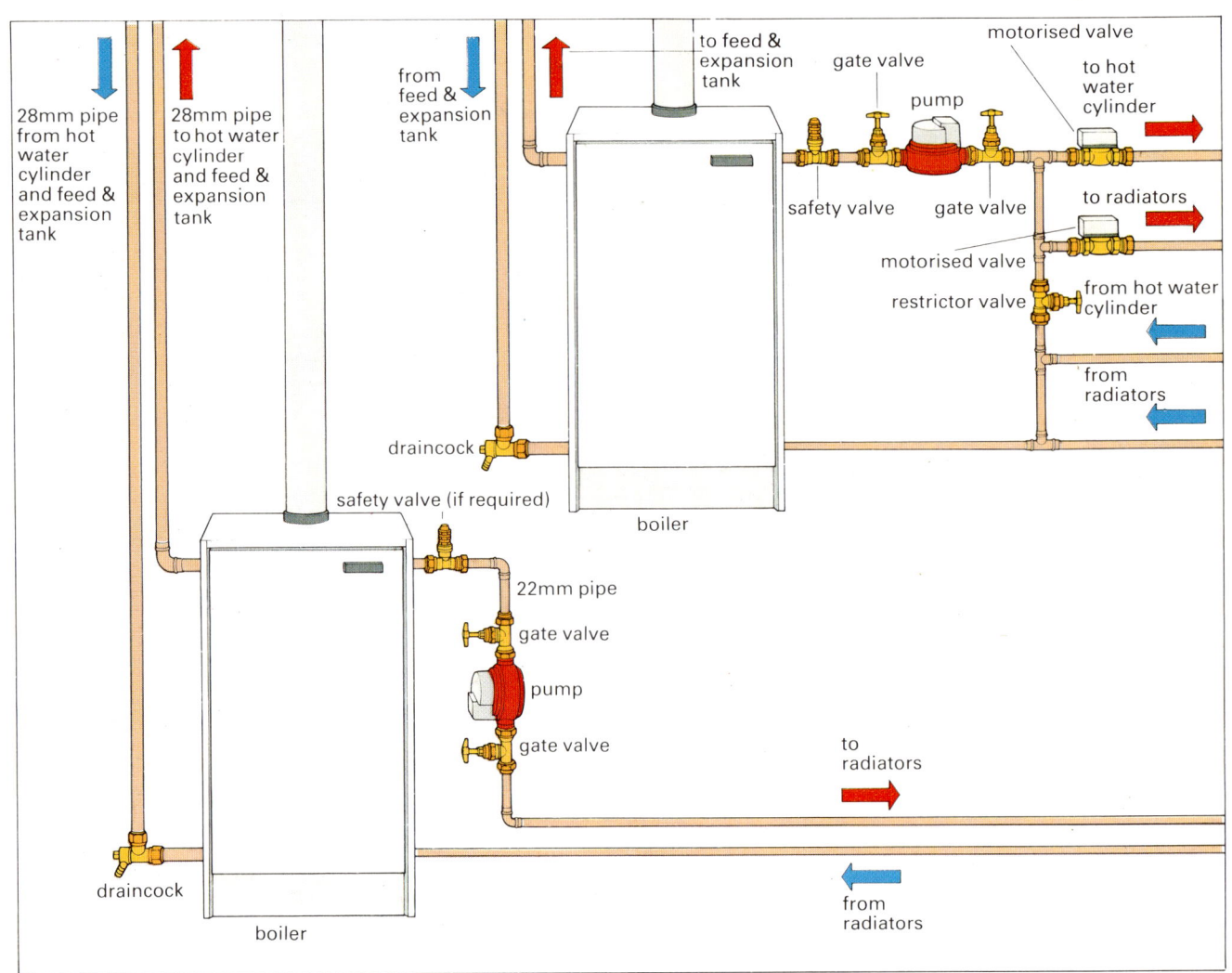

28mm pipe from hot water cylinder and feed & expansion tank

28mm pipe to hot water cylinder and feed & expansion tank

from feed & expansion tank

to feed & expansion tank

gate valve

motorised valve

pump

to hot water cylinder

safety valve

gate valve

to radiators

motorised valve

from hot water cylinder

restrictor valve

from radiators

draincock

boiler

safety valve (if required)

22mm pipe

gate valve

pump

gate valve

to radiators

draincock

boiler

from radiators

With these typical pipework arrangements for the boiler, the top diagram shows fully pumped domestic hot water and central heating circuits, including a bypass circuit which is sometimes required. In the bottom diagram the central heating circuit is pumped, while the domestic hot water circuit is gravity-fed. Here the pumps have been fitted in the flow pipes. If required, they can alternatively be installed into the return pipes.

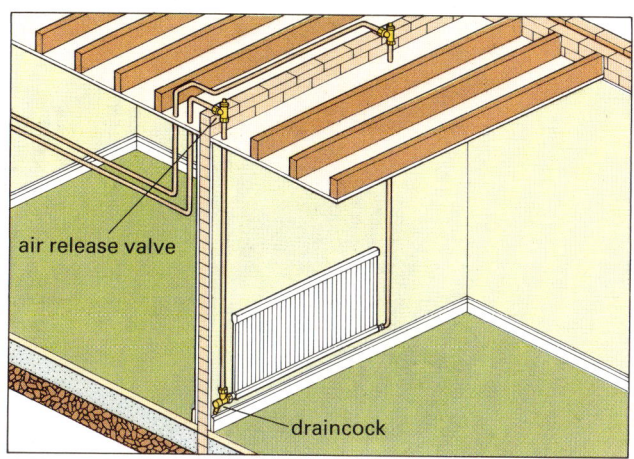

 label: air release valve

draincock

Where there is a possibility of the circuit suffering from the effects of an air-trap, it is important to incorporate an air-release valve at that point. Note in this diagram that a draincock is fitted at the lowest point in the circuit.

There are several advantages to be gained by fitting this system. The materials can work out cheaper, the pipes are easier to manipulate and conceal and the total system contains less water, which in turn requires less heating.

Microbore pipes are also used in sealed systems, which use a pressure vessel to maintain above-atmospheric pressure in the system. This enables you to have higher water temperatures. Such sealed systems are, however, outside the scope of this book.

Because of the narrower bore of the pipes, there is a greater resistance to the flow of water than with small bore pipes. This means that you will need a higher pump head to operate the system effectively.

There are two areas in which the basic system's design normally varies from that of the small bore. Firstly, the standard indirect type of hot water cylinder is not suitable and you will need to install a specific

high recovery cylinder instead. Alternatively you can fit a hot tube type of heating element into the immersion heater boss (see page 80).

The other variation comes in the way in which the water is carried round the system from radiator to radiator. With the small bore system, the radiators are normally on a two-pipe circuit which runs to each radiator. With the microbore system, the resistance to flow in a circuit containing a number of radiators would be too high. Therefore the water is fed to central distribution points called manifolds.

A flow and return manifold is normally provided for each floor in the house. Separate flow and return pipes then supply each radiator on that floor. The pipes feeding the manifolds are, of course, of larger diameter than those feeding the radiators.

The maximum recommended length of microbore pipe from a manifold to each radiator is 7.5 m, giving a total flow and return of 15 m. This means that the manifolds must normally be situated centrally to the required positions of the radiators.

The pipes themselves can be bent quite easily and safely without using a bending spring since their walls are thicker than those of small bore pipes. Where a large bend radius is required, which will give minimal resistance to flow, you can do this by hand across the knee. Where you need a tighter, more consistent bend in a length of pipe, you will need to use a special bending tool that works like a pair of pliers.

Since these pipes are small, the heat loss per metre is considerably less than with small bore pipes and usually you will not have to lag them. Another convenience of using this type of pipe is that it can be channelled into wall plaster or concrete floors, where special PVC-covered pipe is available.

The normal maximum recommended flow velocity of water through the pipes is 1.2 m/sec, which is higher than that for 15 mm pipes. In Table H, you can check the flow-carrying capacity of microbore pipes. The cut-off point in this table is at a velocity of

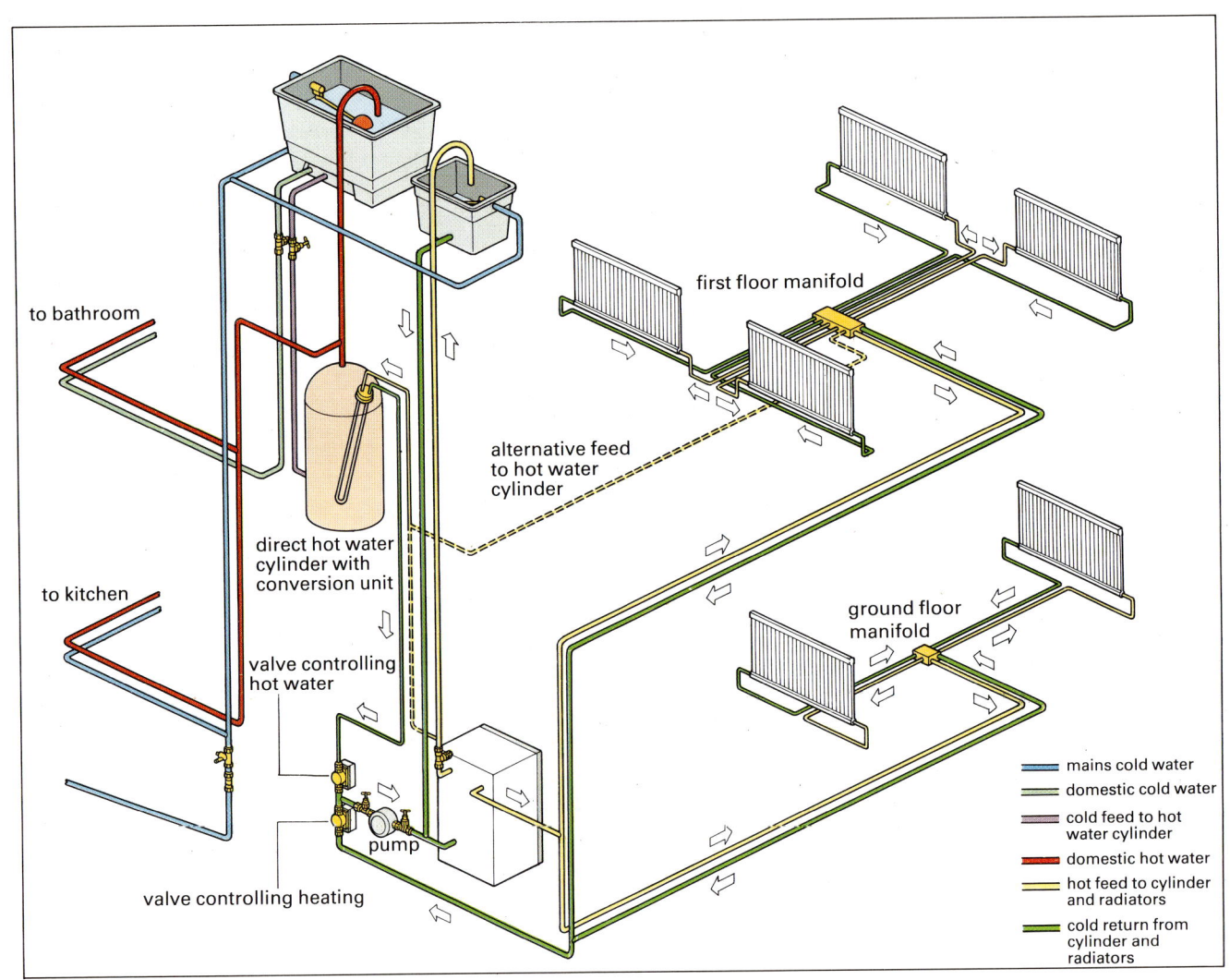

to bathroom

to kitchen

valve controlling
hot water

direct hot water
cylinder with
conversion unit

valve controlling heating

pump

alternative feed
to hot water
cylinder

first floor manifold

ground floor
manifold

mains cold water

domestic cold water

cold feed to hot
water cylinder

domestic hot water

hot feed to cylinder
and radiators

cold return from
cylinder and
radiators

The diagram illustrates a typical system using microbore pipework and fittings. The water in the central heating circuit is fed to the manifolds through the normal small bore pipes. From the manifolds it is then fed to the radiators through the correct size of microbore pipe. Normally one manifold is fitted to feed each floor, depending on the size of the house and the number of radiators. The working pressure for this circuit is greater than for the small bore.

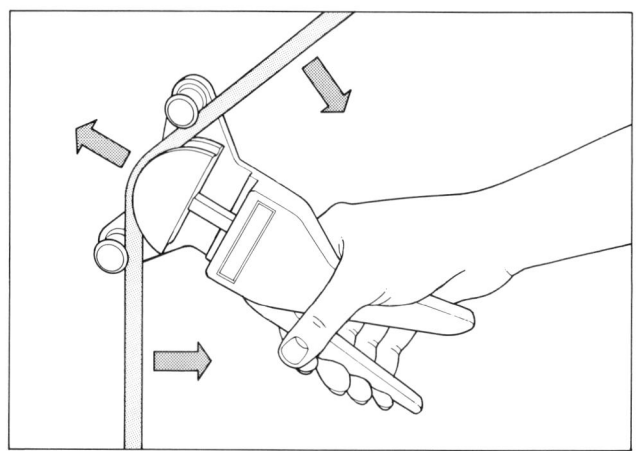

Microbore pipe is much easier to work with and therefore install. Bending is a very simple task, since it can be done using a small hand tool, as shown here.

1.5 m/sec, which is the absolute maximum recommended. The figures in italics are for velocities of between 1.2 and 1.5 m/sec. You can also see that pressure losses for microbore pipes are considerably higher than those for small bore pipes (see Table E on page 63).

Since each loop circuit from the manifolds only feeds one radiator, it is permissible to allow a temperature drop of 15°C from flow to return. This will reduce the pump head required and will also affect the size of the radiator. You should refer to the manufacturer's leaflet for more specific information.

You can make the connections to each radiator in the conventional way using reducing fittings at the valves. Alternatively both connections can be made at one of the lower tappings, using a combined radiator valve. This special valve has two connections – for the flow and return pipes – and is supplied with a nylon tube which is fed into the bottom of the radiator.

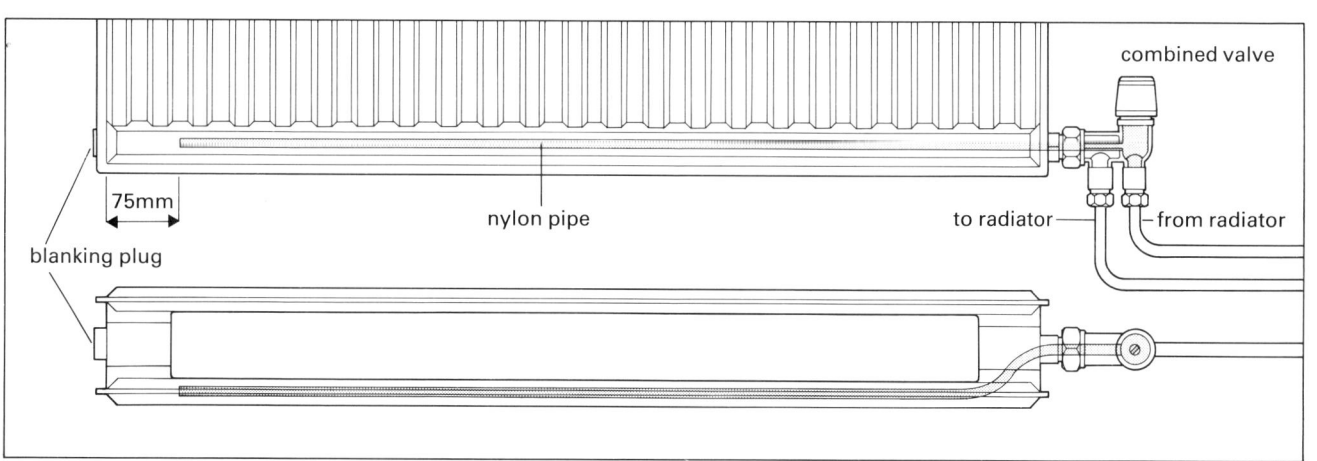

With a microbore circuit, it is possible to connect up the radiators in the conventional way by installing reducing fittings by the control valves. Otherwise you can fit a combined radiator valve in one of the lower tappings.

With this valve there are two connections to take both the flow and return pipes. Supplied with this valve is a length of nylon tube. Feed this along the bottom of the radiator from the flow pipe to ensure overall circulation.

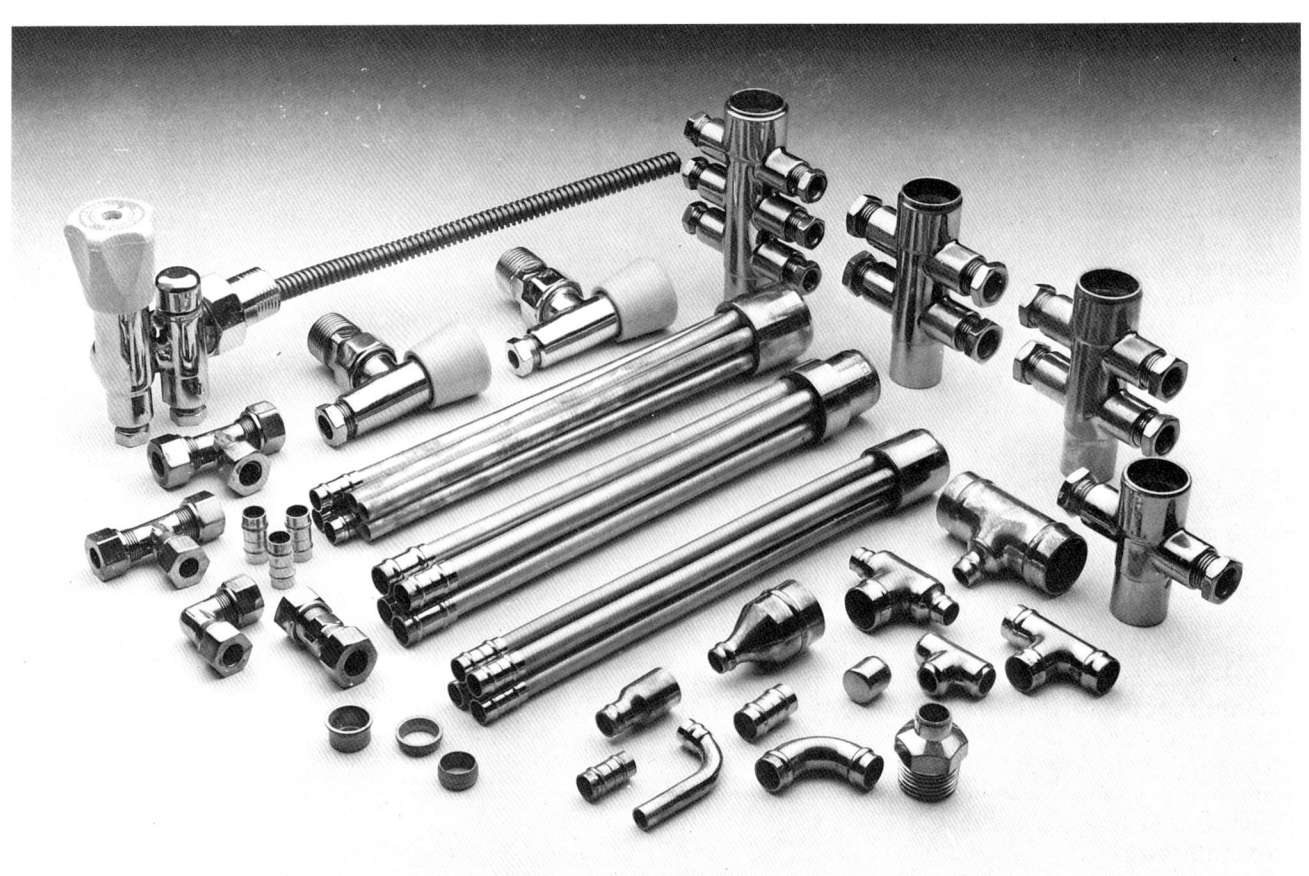

Here you can see a whole range of microbore fittings suitable for use with this type of central heating system.

This includes normal control valves, a single end control valve, manifolds, tees and reducing fittings.

This tube, which should be about 75 mm (3 in) shorter than the length of the radiator, is connected to the flow line into the valve and directs the incoming water to the far end of the radiator. The water then works it way back gradually inside the radiator and out through the return pipe via the valve outlet.

You must take special care when feeding this tube into a double-panel radiator to ensure that the bends necessary in the tube do not cause it to collapse.

7 Getting the system working

Once you have installed the system and plumbed in all the fittings and pipework, you can now fill it with water, make the necessary electrical connections for the controls, commission the system and balance it to ensure that it will then work efficiently and effectively.

Filling the system

Run a check on all the major connections in the system before you turn on the water to fill it. And make sure you first remove the pump to prevent any swarf from the pipes or other debris getting into it and causing damage. You can either use a section of previously prepared piping to replace the pump or fit a section of hosepipe with clips at either end.

Check also that all the valves in the system are open except the radiator air-release valves, which should be closed. As an extra precaution, fit a length of hosepipe, with a clip, on to the draincock at the bottom of the system, in case you have to drain the water off quickly if you discover a leak. Make sure the draincock is closed at this stage.

It is also worth checking that no electrical fittings are in danger of being dripped on should you encounter a leak in the system. A temporary cover of polythene could well prevent a fuse blowing, should this happen.

Assuming that the complete system is dry, tie up the valve on the feed and expansion tank (see diagram) and open the main stopcock to allow water to flow into the domestic hot water circuit. Open all the hot taps and the upstairs cold water taps. Water should flow fairly quickly through the cold taps. As soon as it is

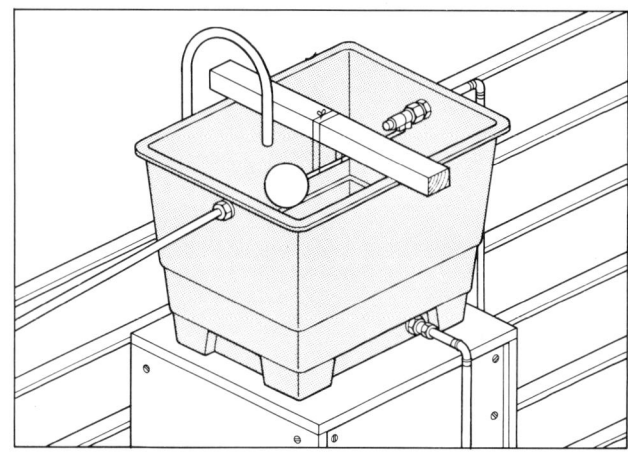

When checking the system for leaks, you will need to hold the valve in the feed and expansion tank closed. You can do this by tying the valve with strong string to a piece of wood and laying this across the sides of the tank.

running freely and there are no signs of dirt or air in the pipes, turn off these taps. Check at this stage that any WC cisterns are filling up as well.

When the hot water cylinder has filled up and water is starting to run up the cylinder vent pipe, it should begin to flow through the hot water taps as well. You can turn off these taps when there are no signs of dirt or air in the pipes.

Now check all the fittings on the hot water cylinder and in the domestic hot water circuit. Go back up into the loft and check the control valve on the cold water storage cistern has closed with the water at the correct level (see page 28).

Then go back to the boiler and open the draincock, making sure before you do so that the hosepipe you connected to it runs outside the house and will not create any problems should water appear. In fact, no water should be present. If it is, this indicates there is a leak in the hot water cylinder calorifier, which will have to be rectified.

Having shut off the draincock, you can now untie the valve in the feed and expansion tank. Let the water run in slowly to start filling the primary circuit.

You will need to bleed any air out of the radiators, which you can do with a special key. Start with those on the ground floor, turning off the air-release valve as soon as water appears. Then turn off the water supply to the feed and expansion tank and check the ground floor fittings for any leaks. Wipe each fitting with a dry cloth to remove any condensation. If you find that any of the compression fittings are leaking, these should respond to further gentle tightening with a spanner. If not, you should be able to remake the joint using some PTFE tape wrapped round the thread.

If a leak occurs in a capillary joint, you will need to drain the system to well below the problem area before heating up the solder in the joint. You may find you can 'top up' the leaking joint with some cored solder, which you should apply to the joint between the fitting and the pipe. If this does not work, you will have to replace the joint and adjoining pipework.

Once you have checked the ground floor, turn on the stopcock again and repeat the bleeding and checking procedure upstairs – including the calorifier joints in the hot water cylinder.

Now go up to the loft again and check that the feed and expansion tank valve has closed with the water at the correct level. If the level of water is too low, you can adjust the arm by turning the nuts on a Garston valve or by moving the arm on to a different notch on a Torbeck valve. If you have the traditional Portsmouth valve, you can make the adjustment by bending the float arm. If the level of water is too high, you will have to drain some off or carefully scoop it out before adjusting the valve. Then turn the stopcock on again.

When you are satisfied that the system is completely watertight, open the draincock by the boiler and let all the water in the system drain away, taking all the dirt and debris with it. Since the feed and expansion tank will now be taking in water again, the system will continue to be flushed through.

After 15 minutes or so, turn off the stopcock controlling the feed and expansion tank and open all the radiator air-release valves, starting upstairs, to ensure the primary circuit drains off completely. Then close the draincock, refit the pump in position, close all the air-release valves and open the stopcock to refill the system.

At this stage i is well worth adding an anti-corrosion inhibitor. This is usually put into the water as the system fills up, but check carefully with the manufacturer's instructions first.

Connecting up the controls

You can now complete all the electrical connections to the controls on the system. Make sure that you follow the manufacturer's instructions correctly and that no controls conflict with each other (see pages 46–54). Where wiring inside the boiler casing – and with some motorised valves – you should use heat-resistant cable. And make sure that all the motorised valves are open and all the thermostats are calling for heat.

One very important point you must check on is that the electricity supply for the whole heating system is controlled by a correctly fused isolator. In most cases this can be achieved using a 13 amp plug in a socket outlet. If you are in any doubt about the wiring up of your control circuit, always call in a qualified electrician to check the work.

Commissioning the system

If you are installing a gas-fired boiler, you should have the gas connected by a fully qualified engineer, who

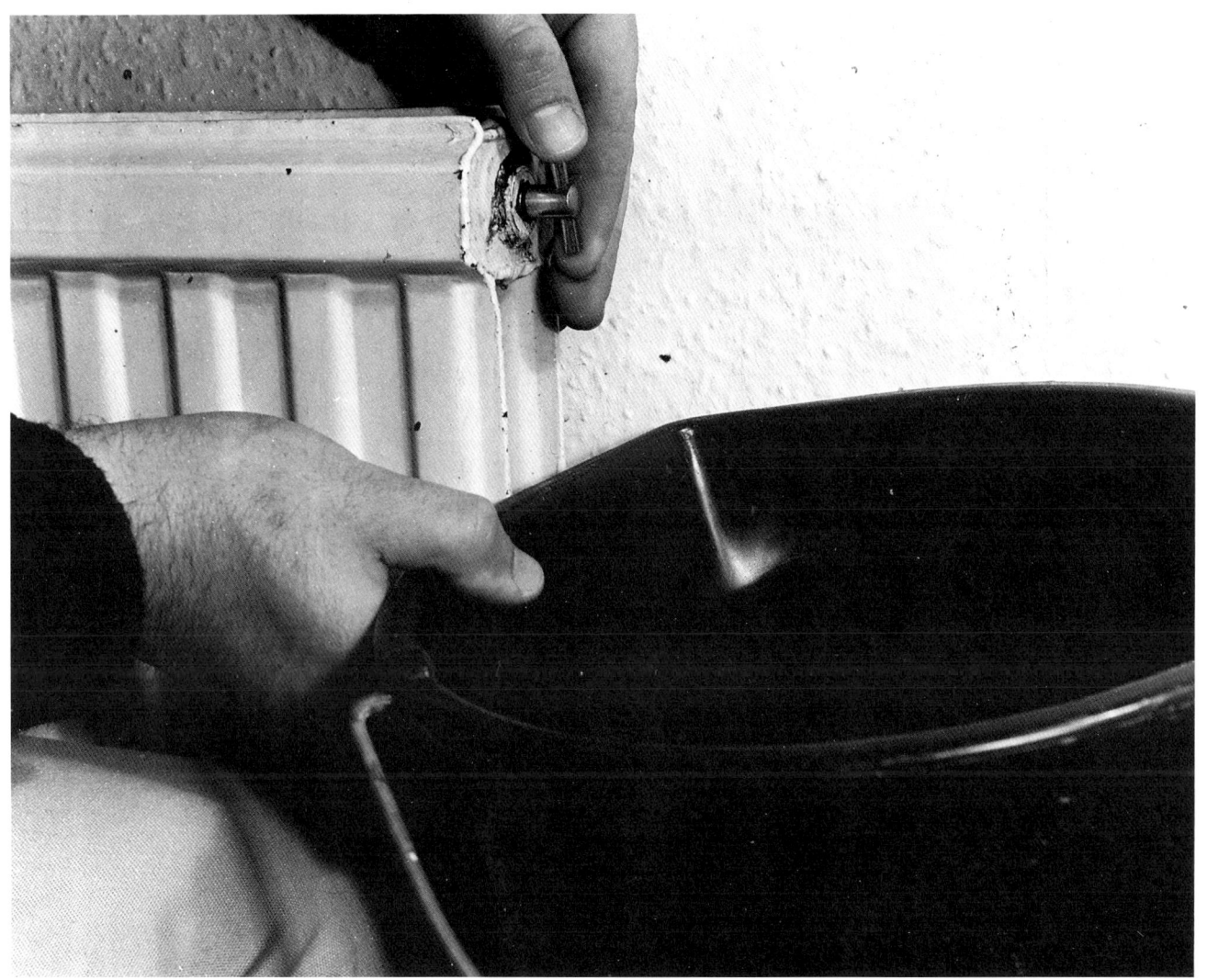

A special key is available to bleed the air out of the central heating system via the air-release valve in each radiator. Starting on the ground floor, test the first radiator until water starts trickling out, then shut off the valve. Although you should not get vast amounts of water gushing out, it is wise to hold a bowl or bucket underneath in case of a mishap. Wipe the valve clean and work round the system, checking each radiator as you go.

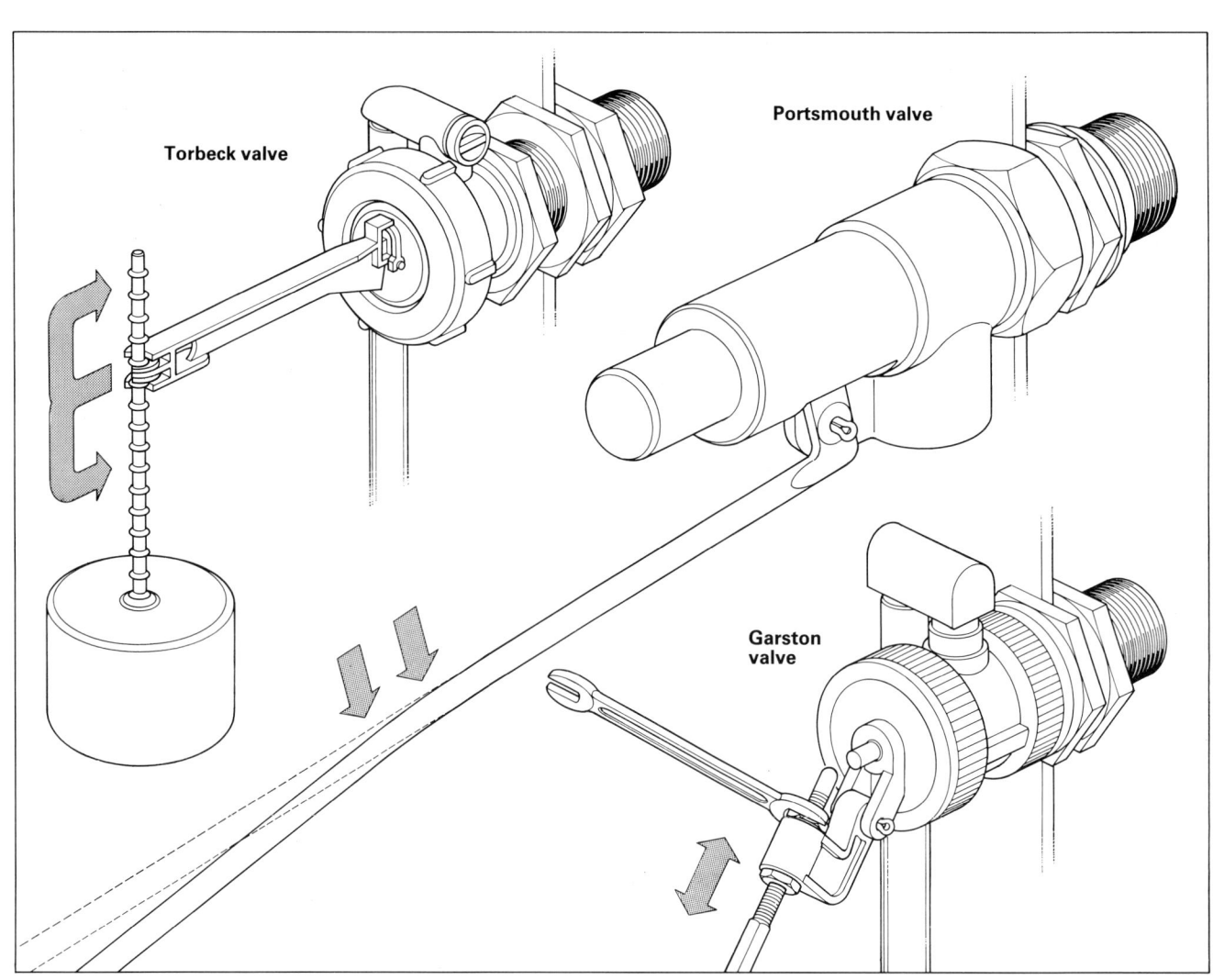

Torbeck valve

Portsmouth valve

Garston valve

There are three basic types of cistern control valves and you should know how to adjust each one. The Torbeck is quite straightforward, since you simply move the float up or down as required. With the Portsmouth the operation is even more basic; you just bend the float arm up or down by hand. Always hold the end near the valve steady, so as not to damage the control. Adjust the Garston by moving the arm in or out of the pivot using the nuts shown.

should also be responsible for the initial pressure setting and the firing of the boiler. If you need any advice on this, contact your local Gas Board. Similar steps should be taken in the case of an oil-fired boiler and here most oil companies or boiler manufacturers will put you in touch with the right engineer.

Set the boiler thermostat to 80°C, the domestic hot water thermostat to 54°C and the room thermostat to 18–21°C. Check that all motorised, radiator, thermostatic radiator and lockshield valves are fully open (although not, of course, the air-release valves). Adjust the pump to the predetermined setting (see pages 82–87) or, if you have not already done the calculation, to a low setting. And, most importantly, check that there is water in the boiler.

Switch on the boiler and let it run for about 15 minutes. During this time, check the security of the fittings on the calorifier circuit if this is gravity-fed. Then switch on the pump. If you have a fully pumped system, the pump will be operating all the time the boiler is on. As the system heats up, recheck all the fittings.

You will have to bleed all the radiators again, but remember to switch off the pump first. This is necessary because as the water heats up it gives off oxygen. Continue this process until the whole system reaches the desired operational temperature and is shut down by the thermostats.

Check in the loft that no water is being discharged through the vent pipe into the feed and expansion tank. If it is, you will have to reduce the pump setting or increase the height of the U bend over the tank.

If you have a gas boiler with a bypass loop (see page 87), you will have to close the bypass valve gradually until the resistance in the bypass loop is slightly greater than that in the index circuit and the water flows around the latter. Make sure, however, that the valve is not fully closed. Wait until the water in the domestic and central heating circuits is at the required operational temperature, when the

thermostats will close both motorised valves.

Check that the pump keeps running for the correct period (according to the manufacturer's instructions) after the boiler has shut down and that there are no boiling noises from the boiler, indicating that the bypass circuit is working correctly. You should check this circuit is functioning properly after you have balanced the system as well.

It is now best to leave the system running for a few days to give it time to settle down before you balance it. This will probably mean that those rooms that are not controlled by thermostats will not be at the correct temperature during this period. Make sure you carry out regular checks on the fittings, especially after the system has cooled down and then reheated after being switched off and on again by the timeswitch.

You can at this stage safely lag all the pipes between floors and in the loft, where necessary. Replace the floorboards – ideally fixing these down with screws so that you know which you had to lift and can move them again easily if required – and make good any damage to the existing decoration.

You will need to fit a lid over the feed and expansion tank, making a hole in it through which to run the U bend of the vent pipe. Lag the tank – and, of course, the cold water storage cistern if you have fitted a new one – with polystyrene panels or blanket insulation. Do not forget to lag the U bend and warning pipe as well; here you can use pipe wrap. Although these pipes should not normally contain any water, you do not want them to freeze up if they do.

Balancing the system

As already mentioned, it normally takes a few days for the new system to settle down and for all the oxygen to be bled off. When this has happened, you are now in a position to balance the system.

If you have installed a fully pumped system and

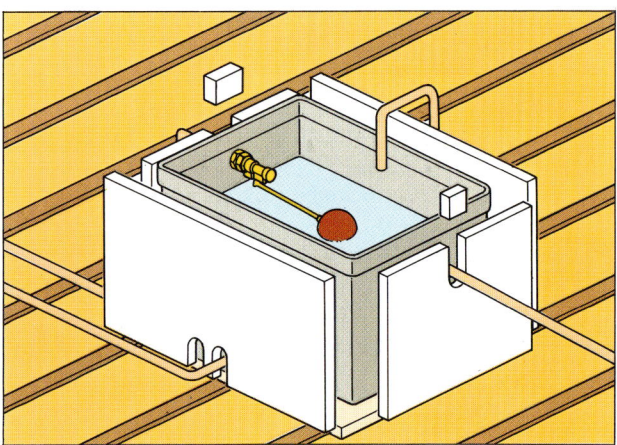

The one area of the loft you should leave uninsulated is the floor under the cold water storage cistern to allow warm air from below to keep it from freezing. Again fit retaining boards if using loose-fill insulation.

fitted a lockshield valve in the calorifier circuit, this valve is use to vary the amount of water being pumped around the circuit. It could be that the calorifier circuit has a very low flow resistance. If this is the case, then the primary water will flow round this circuit in preference to the central heating circuit. This will result in the domestic hot water heating up quickly, while the rest of the house will remain cold until the hot water cylinder thermostat shuts off the motorised valve. Only then will the central heating circuit receive all the boiler's output.

This situation may be acceptable to you. But if you want the house to heat up as well as the domestic hot water when the boiler first switches on, then you should screw down the lockshield valve to reduce the flow of water to the calorifier. As a general guide, the domestic hot water should reach the desired operational temperature in about two hours.

When trying to balance the system, you will need a pair of clip-on thermometers to fix on to the flow and return pipes to the radiators. Ideally you should balance the system while the outside temperature is about −1°C, although this is not critical and, of course, not always practical.

Open all thermostatic radiator valves on the index circuit fully and check to see which radiators are hottest. You can normally do this by hand. Now go back and screw down slightly the lockshield valves on the hottest radiators and leave the system alone for an hour or so to give it time to settle down. Clip the thermometers on to the flow and return pipes of all the radiators in turn on the same circuit, leaving them on for a few minutes, and note the difference in the readings.

Keep adjusting the lockshield valves on the offending radiators until the temperature difference readings are consistent on each radiator. To increase the difference between the flow and return pipes, you need to screw down the lockshield valves. When the readings are the same on all radiators, the circuit is balanced.

Repeat this procedure for any other central heating circuits in the house to balance the whole system. The difference between the flow and return pipes should be that for which the circuit was designed.

For small bore circuits, this should be 10–11°C.

For microbore circuits with convector radiators, different figures may apply and you will need to check these out with the manufacturer or supplier. If, after you have balanced the system, the temperature difference is too high, you should increase the pump setting – and reduce it if the difference is too low.

Index

air-release valve, radiator 31, 100, 107, 108, 111
airing cupboard 45, 94
airlocks 100, 108
anthracite 15, 16, 33
anti-corrosion inhibitor 108
ascending spray bidet 99
ash 7, 16, 33

back boiler 16, 33, 36, 38
background central heating 7, 47
balancing the system 111-113
bathroom 20, 56, 69–71
bedrooms 56
bedsits 56
bending machine 15, 22, 24
bending springs 22–25, 103
blowtorch 24, 25–26
boiler 7, 13, 14, 15–19, 33–41, 46, 50, 53, 56, 76, 81,
    86, 87, 111, 112
    calculating size of 81
    Economy 7 18, 19
    fitting 92–94
    freestanding 40, 92
    gas-fired 18, 38–39, 108–109, 111
    gravity-fed 33
    isolated 40–41
    oil-fired 18, 36–37
    plumbing in of 100–101
    solid fuel 15, 16, 33–36, 71
    servicing of 37
    wall-mounted 39, 40, 54
brass fittings 9, 27
British Standards 9
bypass pipework 87, 111

calorifier 13, 15, 47, 50, 88, 94, 100, 108, 112
    gravity-fed 15, 16
calorifier circuit 81, 111, 112
capillary fittings 22, 25, 26, 27
capillary joint 25, 26, 27
carpets 11
cavity wall insulation 8, 64
ceilings 59
chimneys 9, 38, 41, 42–43
cistern valves 31, 107
clinker 16
coal 15, 33
coal bunker 21
Coal Utilisation Council 16
coke 15
cold feed pipe, boiler 82, 100
cold water storage cistern 9, 15, 28, 45, 107
    fitting 94
    plumbing in 99–100
compression fittings 25, 26, 27, 29, 107
compression joint 26–29
compression nuts 26–27, 29
compression ring 27, 29
condensation 56
    of flue gases 41
control circuit, wiring up 108
control systems 46–54
convector heaters 18
conversion of units 9–10
cost comparisons
    central heating systems 11
    heating fuels 7–8, 20–21
curtains 72
cylinder vent pipe 82, 107

Department of Energy 33
dezincified fittings 27, 29
dining room 56
domestic hot water 12, 14, 16, 19, 33, 45, 46, 47, 50,
    56, 87, 112

calculating heat for 64–66
calorifier 47
circuit *see* secondary circuit
gravity-fed system 16, 33, 79, 82, 86
pumped system 82
doors 59
double glazing 9, 56, 59
draincock 31, 107–108
draughts 9, 18, 51, 56, 71
draughtproofing 8
ducted warm air systems 7, 12

Economy 7 tariff 8, 13
elbow fitting 22
electrical connections 54, 108
electrical safety 54, 107
electricity 8, 18, 20
boiler (Economy 7) 18, 19
off-peak 7, 8, 11, 12, 13
Electricity Board 7, 13, 19, 56
expansion tank *see* feed and expansion tank

fan convector, kick-space 74–75
feed and expansion tank, 14, 15, 45, 86, 94, 95, 107, 108, 111
fitting 94
plumbing in 100
valve 108
filling the system 107
fittings 22–32
flame-sensing devices 38–39
floorboards, lifting 95–97
floors 11, 12, 19, 59
concrete 74, 76, 103
wooden 76, 97
flow velocity 103–105
flue lining 41–44
flues 16, 18, 33, 36, 37, 38, 41–45
balanced 18, 37, 39–40, 92
external 9, 33, 36, 44

open 40
frost thermostat 20, 41, 51
fuel economy 19, 20, 37, 38
fuels, heating 7–8, 15–18
cost comparisons of 20–21
gas 7
liquified petroleum gas 37
oil 8, 36
smokeless 33, 34, 35
solid 7, 15–16, 33, 34–35
furniture 73

Garston valve 31, 99, 108
gas 7–8, 21, 36, 38
boiler 18, 38–39, 108–109, 111
cylinders 8, 37
piped 8
propane 18, 37
Gas Board 37, 38, 45, 109
gas combination units 45
gas connection 18, 21
gas fire with back boiler 38
gas supply piping 92
gate valves 28, 47, 86, 99, 100
gravity-fed systems 33, 40, 79
pipes for 22
gunmetal fittings 27

hacksaw 22
hall 56
header tank 14
heat, surplus 33, 71
heat emission, calculations of 74, 75, 76, 78
heat exchangers 12, 68, 69
heat flow rate 76–79
heat load, total 79, 81
heat losses or gain 56, 59–64, 65, 72, 81, 103
heat reflecting panels 72, 92
heat requirements 56–66, 76, 78, 81
heat resistant cable 108

heat transmittance *see* U valve
heating cost comparisons 8, 20–21
heating engineer 37, 111
hot water cylinder 8, 15, 28, 45, 53, 87, 88, 103, 108
    calorifier 15, 79
    fitting 94
    plumbing in 100
    thermostat 50, 87, 110, 112

immersion heater 13, 18, 19, 88
    removal of 88
Imperial measurements 9–10, 22
insulation 8–9
    cavity wall 8, 64
    grants for 9
    hot water cylinder 8
    loft 8, 56, 64
    see also lagging

jointing compound 27
joists, notching 97

kitchen 20, 56, 69, 76
kitchen sink 15

lagging 66, 75, 76, 78, 103, 111
landing 56
leaks 107, 108
liquified petroleum gas (LPG) 37
lockshield valves 31, 76, 111, 112
loft insulation 8, 56, 64
lounge 56, 75

mains water supply 9, 15
manifolds 15, 103, 105
microbore system 15, 101–103, 105
minibore system *see* microbore system
motorised valves 50, 51, 53–54, 87, 108, 111

off-peak electricity 7, 8, 11, 12, 13
oil 8, 18, 20–21, 36
oil filter 36
oil storage tank 21, 36, 37
oil-filled radiators 18
oil-fired boiler 18
olive 27
on-off switch 47
open fire (with back boiler) 36

pipe circuit, designing 76–79
pipe clips 25, 26
pipe cutting 22
pipe fittings 25–32
pipe runs 15, 73, 94, 101
pipe sizes, calculating 78–81
pipe stop 27, 29
pipes 22–32, 97–106
    bending 22–25, 103
    cutting 22
    copper 15, 22, 75
    galvanised steel 22
    holding 25
    PVC-covered 103
    stainless steel 14, 22
planning permission 9, 36, 44
planning system 56–81
plumbing in central heating 99–106
Portsmouth valve 31, 108
primary circuit 14–15, 47, 50, 79, 81, 87, 94, 100, 108
programmers 48–50, 51, 54
propane gas 18, 37
pump, central heating 15, 18, 28, 46, 47, 50, 53, 54, 82–86, 87, 107, 111
    bypass 100–101
    positioning of 82–86
    pressure 76
    setting 113
    size, calculation of 76, 79

radiant heaters 18
radiator air-release valves 31, 100, 107, 108, 111
radiator shelf 73, 75
radiator shut-off valves 31, 33
radiators 13, 14, 15, 20, 33, 76, 103, 105–106, 108, 111, 112–113
    convector 47, 69, 75, 79, 113
    double panel 106
    fitting 88–92
    panel 66, 75
    plumbing in 101
    siting 71–74
    size, calculation of 75, 76
    skirting 14, 47, 66–69, 75, 76, 92
reduction fittings 22, 105
reflective aluminium foil panel 72, 92
regulations
    asbestos 44
    planning authority 9
    smoke control 16
    water authority 9
resistance to flow 79, 101, 103
room temperature 56, 75, 76
    control of 20, 46, 47, 50–54
rotary pipe cutter 22

safety valves 82
sealed system 13–14, 101
secondary circuit 15, 50, 79, 94, 107, 111
shower unit 99
skirting radiators 14, 47, 66–69, 75, 76, 92
    combined with panel radiators 75
    fitting 92
small bore system 14–15, 103, 105, 113
    pipes for 22
soft water areas 9, 27
soldering joints 25, 26–27, 28, 108
solid fuel 7, 15–16, 33, 34–35
    boiler 15, 16, 33–36
    storage of 16, 21, 33

Solid Fuel Advisory Service 36, 81
stopcock, main 107, 108
storage radiators 7, 12–13
    thermostat control of 12–13
swarf 22, 107

tap connectors 22
thermostats 50–53, 54, 69, 76, 108, 111
    boiler 47, 50, 75, 87, 110
    central heating circuit 50, 87
    frost 20, 41
    hot water cylinder 50, 87, 110, 112
    remote 69
    room 50–51, 86, 110
thermostatic control switches 20
thermostatic damper 16
thermostatic radiator valves (TRV) 20, 31, 51–53
    remote reading 53
timeswitches 19–20, 46–47, 48–50, 87
tools 22
Torbeck valve 31, 99, 108
towel rail, bathroom 69–71, 76

U values 59, 64
U bend 15, 82, 86, 99, 100, 111
underfloor heating 7, 11–12
union connectors 100

valves 28–32, 82, 100, 105–106, 107
    air-release, radiator 31, 100, 107, 108, 111
    cistern 31, 107
    draincock 31, 107–108
    Garston 31, 99, 108
    gate 28, 47, 86, 99, 100
    lockshield 31, 76, 111, 112
    motorised 50, 51, 53–54, 87, 108, 111
    Portsmouth 31, 108
    radiator shut-off 31
    thermostatic radiator 20, 31, 51–53, 111, 112
    Torbeck 31, 99, 108

vent pipe, boiler 15, 82, 86, 99, 100
ventilation 8, 18, 33, 36, 56, 59
   heat required for 56, 59, 64

wall plaster 76, 103
walls 72, 73
warning pipe 94, 99, 100, 111

water, soft 9, 27
water outlets 15
water supply, mains 9, 15
wet systems 7, 13
windows 9, 39, 64, 71, 72, 74, 75
   double glazed 9, 56, 59
wood-burning stoves 16, 18